AFRICAN ISSUES

Series Editors Alex de Waal & Stephen Ellis

Killing for Conservation Wildlife Policy in Zimbabwe

ROSALEEN DUFFY

Angola From Afro-Stalinism to Petro-Capitalism

TONY HODGES

Congo-Paris Transnational Traders on the Margins of the Law

JANET MACGAFFEY & RÉMY BAZENGUISSA-GANGA

Africa Works Disorder as Political Instrument

PATRICK CHABAL & JEAN-PASCAL DALOZ

The Criminalization of the State in Africa

JEAN-FRANÇOIS BAYART, STEPHEN ELLIS
& BEATRICE HIBOU

Famine Crimes Politics & the Disaster Relief Industry in Africa

ALEX DE WAAL

Above titles
Published in the United States & Canada
by Indiana University Press

Peace without Profit How the IMF Blocks Rebuilding in Mozambique

JOSEPH HANLON

The Lie of the Land Challenging Received Wisdom on the African Environment

Edited by
MELISSA LEACH & ROBIN MEARNS

Fighting for the Rain Forest War, Youth & Resources in Sierra Leone

PAUL RICHARDS

Above titles
Published in the United States & Canada
by Heinemann (N.H.)

AFRICAN ISSUES

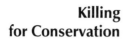

**Killing
for Conservation**

Wildlife Policy
in Zimbabwe

Killing for Conservation

Wildlife Policy in Zimbabwe

ROSALEEN DUFFY
Lecturer in Politics
Lancaster University

The International
African Institute

in association with

JAMES CURREY
Oxford

INDIANA UNIVERSITY PRESS
Bloomington & Indianapolis

WEAVER
Harare

#4476 8909

The International
African Institute
in association with
Weaver Press
PO Box A1922
Avondale, Harare
.

James Currey
73 Botley Road
Oxford OX2 0BS
&
Indiana University Press
601 North Morton Street
Bloomington
Indiana 47404
(North America)

© Rosaleen Duffy 2000

First published 2000

1 2 3 4 5 04 03 02 01 00

British Library Cataloguing in Publication Data
Duffy, Rosaleen
Killing for conservation: wildlife policy in Zimbabwe. –
(African issues)
1. Wildlife conservation – Zimbabwe
I. Title
639.9'096891

ISBN 0-85255-846-5 (James Currey paper)
ISBN 0-85255-845-7 (James Currey cloth)

Library of Congress Cataloguing-in-Publication Data
Duffy, Rosaleen.
Killing for conservation: wildlife policy in Zimbabwe/Rosaleen Duffy.
p. cm. – (African issues)
Includes bibliographical references (p.) and index.
ISBN 0-253-33915-4 (cl: alk. paper) – ISBN 0-253-21454-8 (pa: alk. paper)
1. Wildlife conservation–Government policy–Zimbabwe. I. Title. II. Series.

QL84.6.Z55 D84 2000
333.95'416'096891–dc21   00-061370

Typeset by
Saxon Graphics Ltd, Derby
in 9/11 Melior with Optima display
Printed and bound in Great Britain by
Woolnough, Irthlingborough

CONTENTS

ABBREVIATIONS ix

ACKNOWLEDGEMENTS xiii

INTRODUCTION

The African Environment & Green Ideology 1

1

Depoliticising & Repoliticising Wildlife Conservation & Sustainability 9

2

Individual Power & Personal Patronage in Conservation Politics 21

3

The Use of State Force in Anti-poaching & in Poaching 43

4

Privatising Wildlife Conservation 67

5

Community Participation in Campfire as Political Legitimation 89

6

Buying Influence? The Politics of Donor & NGO Involvement 113

7

The International Politics of the Rhino Horn & Ivory Trade 141

CONCLUSION 173

BIBLIOGRAPHY 179

INDEX 205

AERSG	African Elephant and Rhino Specialist Group
AESG	African Elephant Specialist Group
ART	Africa Resources Trust
AWF	African Wildlife Foundation
BSAC	British South Africa Company
Campfire	Communal Areas Management Programme for Indigenous Resources
CASS	Centre for Applied Social Science (University of Zimbabwe)
Cde.	Comrade
CFTW	Care for the Wild
CFU	Commercial Farmers Union
CIO	Central Intelligence Organisation
CITES	Convention on International Trade in Endangered Species
CND	Campaign for Nuclear Disarmament
CSI	Chief of Staff (Intelligence)
DFID	Department for International Development
DNPWLM	Department of National Parks and Wildlife Management
DTZ	Development Trust of Zimbabwe
EC	European Community
EIA	Environmental Investigation Agency
EPO	exclusive prospecting order
ESAP	Economic Structural Adjustment Programme
EU	European Union
Frelimo	Frente de Libertação de Moçambique
GEF	Global Environment Facility
GPS	Global Positioning System
GTZ	Deutsche Gesellschaft für Technische Zusammenarbeit
ICA	intensive conservation area
IFAW	International Fund for Animal Welfare
IFRA	Institut Français des Recherches Africaines

IMF	International Monetary Fund
INTERPOL	International Criminal Police Organization
IPZ	Intensive Protection Zone
ITN	Independent Television News
ITRG	Ivory Trade Review Group
IUCN	International Union for the Conservation of Nature
LSCF	large-scale commercial farming
MAPS	multispecies animal production systems
MLGRUD	Ministry of Local Government, Rural and Urban Development
MMD	Movement for Multiparty Democracy
NCS	National Conservation Strategy
NGOs	non-governmental organisations
NR	natural region
ODA	Overseas Development Administration
OXFAM	Oxford Committee for Famine Relief
PAC	Problem Animal Control
PF-ZAPU	The Patriotic Front-Zimbabwe African People's Union
PVO	private voluntary organisation
Renamo	Mozambique National Resistance
RNFU	Rhodesian National Farmers Union
SACIM	Southern African Centre for Ivory Marketing
SACWM	Southern African Centre for Wildlife Management
SADC	Southern African Development Community
SADF	South African Defence Force
SAGE	Southern African Group for the Environment
SAVE	Save African Endangered Wildlife
SAS	Special Air Service
SPFE	Society for the Preservation of the (Wild) Fauna of the Empire
SVC	Save Valley Conservancy
TCM	traditional Chinese medicine
Traffic	Trade Records Analysis of Flora and Fauna in International Commerce
UDI	unilateral declaration of independence
UN	United Nations
UNCED	United Nations Conference on Environment and Development
UNDP	United Nations Development Programme
UNEP	United Nations Environment Programme
UNIP	United National Independence Party
UNITA	União Nacional para a Independência Total de Angola
UNOSOM	United Nations Operation in Somalia
USAID	United States Agency for International Development
UTC	United Touring Company
VIDCO	Village Development Committee

WINDFALL	Wildlife Industries for All
WISDOM	Wildlife in Sustainable Development of Mankind
WMS	Wildlife Management Services
WPA	Wildlife Producers Association
WTMU	Wildlife Trade Monitoring Unit
WTO	World Tourism Organisation
WWF	World Wide Fund for Nature/World Wildlife Fund
Zimtrust	Zimbabwe Trust
ZANU	Zimbabwe African National Union
ZANU-PF	Zimbabwe African National Union–Patriotic Front
ZAPU	Zimbabwe African People's Union
ZATSO	Zimbabwe Association of Tour and Safari Operators
ZHA	Zimbabwe Hunters Association
Zimnet	Zimbabwe National Environmental Trust
ZNA	Zimbabwe National Army
ZNCT	Zimbabwe National Conservation Trust
ZPG	zero population growth
ZTDC	Zimbabwe Tourism Development Corporation

ACKNOWLEDGEMENTS

I must thank the University of Lancaster and the Department of Politics and International Relations for providing the funding which allowed me to complete my doctoral thesis (on which this book is based). I would like especially to thank Professor Christopher Clapham, who has been a constant source of support; he read numerous drafts of this book and he dutifully nods each time I come out with a new tall story from my fieldwork in Zimbabwe. Likewise, Dr Stephen Ellis has invested a great deal of time and effort during the lengthy process of getting this book into shape, and he too has been very supportive of my efforts.

In Zimbabwe I was reliant on the hospitality of Peter, Melissa and David and they also allowed me to mull over thoughts about the information I had collected with them – sometimes on a more than daily basis. There are a number of people in Zimbabwe I would like to thank for their courage in providing me with politically sensitive information, and for that reason they cannot be named here. The staff of the Parks Department, the Parks library, Modus Publications and the University of Zimbabwe were also vital players in the completion of this book.

The African Environment & Green Ideology

Wildlife conservation in Africa is an issue that has captured the public imagination in the industrialised world. Wildlife documentaries are watched with keen interest, children's cartoons endow the animal kingdom with human-like characteristics, and international conservation organisations run highly public campaigns to save African big mammals, such as elephants and rhinos, from greedy poachers. The commonly accepted image is that African wildlife is under threat from African people, and so African wildlife has to be saved and preserved in the interests of global environmental good. Wildlife conservation in Africa is widely discussed in newspapers, magazines and television programmes in the industrialised world. However, little attention is paid to what conservation actually means and, in the end, a greatly simplified view of a very complex issue is communicated to readers and viewers. Broadly, conservation and preservation constitute different approaches to wildlife policy, since preservation is more concerned with protecting a habitat, an animal or a resource from any use at all, while conservation allows some use but ensures against extinction or overuse. Since sustainable development became the fashionable buzz-words, this problem has been compounded. The lack of a precise definition or even a set of guiding principles for sustainable development also feeds into the depoliticising rhetoric of wildlife conservation as a global environmental 'good'. As Leach and Mearns argue, environmental narratives serve to standardise, package and label environmental problems so that they appear to be universally applicable and to justify equally standardised off-the-shelf solutions (Leach and Mearns, 1996: 8). Goodin argues that for the purposes of public policy debates, subtle ethical doctrines are translated into generalised principles that ultimately lose their impact and importance (Goodin, 1983). Rather like the old adage that sport and politics don't mix, the slogan is equally misplaced when applied to the environmental movement leading to claims that the environment is above politics or does not mix well with politics. Just as sporting events

1

that are held in the face of calls for political and economic boycotts act as clear political expressions of a level of international acceptance, the environment is closely bound up with politics.

The way that wildlife is presented – in peril and able to be conserved by politically neutral conservation strategies – masks the inherently political nature of conservation policy-making at the local, national and international levels. Environmental issues and conservation policies are most often presented as non-political and as neutral or scientific prescriptions, but an analysis of how science is constructed and the differing ways in which scientific knowledge is used has been lacking (Wynne, 1992: 111–27; Gorz, 1980: 17–20). This means that the rhetoric of conservation can be used to depoliticise a highly complex set of ethical, political, social and economic dilemmas. Concepts and representations of the environment inevitably have a critical impact on how ecological issues are viewed, and this in turn determines the policy responses designed to manage the environment.

For the purposes of this book, an ideology is used to mean a system of ideas to explain, justify and legitimate actions and goals (Olsen et al., 1992: 13–22). The rhetoric of conservation can be utilised by the various interest groups involved to support the ways they want to appropriate or allocate key resources, such as land or lucrative tourism deals. As a result, wildlife conservation in Africa has persistently been an area for heated political debate in the local, national and international political arenas. Environmental policy-making can be presented as part of a legitimating ideology for the political objectives put forward by states, non-governmental organisations (NGOs) and international organisations. This book will examine the ways that different interest groups at local, national and international levels mobilise and promote various ideologies of conservation, from strict preservation to sustainable utilisation, to justify and legitimate their particular agendas in terms of use, allocation and appropriation of potentially lucrative wildlife resources in Africa. In so doing, they use the depoliticising rhetoric of conservation as uncontested science to escape the complex ethical and political considerations that lie at the heart of policies that ultimately result in land and other resources being targeted for wildlife conservation.

While the main focus of this book is the international politics of Zimbabwe's wildlife policy, the debates surrounding the politically controversial policy agenda of wildlife utilisation indicate the wider issues in the realm of global environmental politics. It is important to pay attention to the global debates surrounding notions of sustainable development, which have become particularly important for developing countries. Wildlife conservation policy is often discussed with reference to the broader debates of sustainable development. In particular, there are two key areas of Zimbabwe's wildlife policy that bring the disputes over conservation and sustainable development into sharpest focus, and they are the Communal Areas Management Programme for Indigenous

Resources (Campfire) and the Convention on International Trade in Endangered Species (CITES). Campfire is a world-famous programme aimed at integrating wildlife conservation with rural development by allowing subsistence farmers to use the wildlife found in their area to generate revenue for community development schemes. For example, a quota of elephants can be sold to a safari operator, and a portion of the revenue derived from hunting those elephants is passed to the community. The position adopted by the Zimbabwe Government and a number of local conservation organisations towards CITES is equally controversial since it is the stated aim of the government to overturn the international ban on the ivory and rhino horn trades. Campfire and the trade in products from endangered species are at the forefront of global debates about the efficacy and importance of welding together community development and wildlife conservation, and about the rights of developing countries to use their natural resources as revenue generators.

The question of whether environmental care and development are compatible processes has been a focus of debates on sustainable development in the South. It is usually assumed by the environmental movement that economic development, export-oriented growth and industrialisation are forms of development that are not compatible with conservation of the environment. For some analysts, the drive for development and environmental conservation are contradictory. This stems from the belief that the natural environment must be exploited for development, and that this will probably lead to the kind of degradation and environmental change witnessed by the industrialised world. It also includes the 'limits to growth' debate, which focuses on the argument that world resources are limited and that development processes necessarily deplete them until they are exhausted (Hardin, 1968: 1243–8; Meadows, 1972). The United Nations Conference on Environment and Development (UNCED) held in Rio de Janeiro in 1992 represented the divided views on environment and development. Leaders from southern states argued that environmental conservation was a luxury they could not afford since it constituted an impediment to development. Industrialised states, on the other hand, were keen to point out that industrialisation and development in the South constitute a global environmental threat. As a result, countries of the South were to be persuaded to curb industrialisation in favour of the global environmental good. The South perceived the northern position as one of double standards, claiming that the First World contributed to most of the current global environmental problems and was least willing to provide finance to tackle them.

It is questionable whether the twin concerns of environment and development are irreconcilable. Certainly current patterns of development that emphasise economic growth at the expense of social and political development have damaged the environment, even though it is in the South that environmental care is crucial. While in the North environmental concern centres around a clean-up of the adverse by-products of industrialisation, in the South environmental degradation is very much a survival issue. For countries that rely on exporting crops and other raw materials, degradation of the natural resource base can affect crop yields and, consequently, national income. At the other end of the spectrum, environmental damage affects the rural dweller who relies on water, land and wood. A number of commentators have viewed poverty, inequality and environmental degradation as a self-reinforcing triangle leading to greater and greater poverty (Bruntland, 1987; Redclift, 1987; Chambers, 1988; Johnson and Nurick, 1995). However, for the industrialised nations, Woodhouse suggests that there are two dimensions of concern for sustainable development in the South: the fear that industrialisation will increase global pollution and contribute to global climate change, and also the 'Third World' element to planetary problems, such as population growth (Woodhouse, 1992: 97–116). The concept of sustainability has become so important to the North that environmental conditionality has become a feature of some loans and aid packages, even though there is no agreed definition of sustainability.

Implementation of sustainable development is problematic because of the lack of any clear and coherent statement of the theory. The term 'sustainable development' has been used to describe the many different theories and development projects that include a concern for the environment. Generally, the now internationally famous definition of the Bruntland Commission has come to be accepted as the guiding principle of sustainable development: 'It meets the needs of the present without compromising the ability of future generations to meet their own needs.' (Bruntland, 1987: 8). The Bruntland Commission envisaged a future that would be more economically and socially prosperous, just and environmentally sustainable. It balances satisfaction of basic need within certain (unspecified and unquantified) limits. The Bruntland definition has been criticised as unworkable and ultimately meaningless. In particular, it did not provide guidance on what constitutes needs for present and future generations (Beckerman, 1994: 191–209), and it failed to address how increased economic growth could be made compatible with environmental protection.

The difficulty with notions of sustainability is that the apparent simplicity of the term and lack of precise definition have meant that its vagueness allows the word to be applied to almost anything to do with schemes that claim to conserve the environment. As a result, various interest groups have tended to exploit this ambiguity, captured the term and used it to underpin their own interests by attaching the label to any project or policy that has environmental components (O'Riordan, 1988: 29–51). The notion of sustainability has become so fashionable amongst governments, the private sector, NGOs and donors that its adoption by practically every conceivable type of institution has ensured that it has become a meaningless tag. Sustainability is promoted by neoliberal financial institutions, such as the World Bank, and at the same time it is touted by radical animal rights pressure groups. It stretches the imagination that the World Bank and animal rights activists constitute genuine bedfellows. The reason very different sets of interest groups come to talk about sustainability is that they subscribe to very different notions of what sustainable development actually means, without detailing the differences between them; rather, sustainable development is promoted as a common good, without an analysis of its precise definition. The notion of environmental sustainability appeals to a wide spectrum of interest groups precisely because it is a catch-all phrase, but in order to examine conservation strategies in sub-Saharan Africa it is important to unpick the diverging strands of green political thought that are encompassed in the slogan 'sustainable development'. The increasing sophistication of environmental thought is reflected in the variety of views that are collectively referred to as the green or environmental perspective.

First, blue-greens can be defined as the conservative or right-wing end of the environmental spectrum. Blue-greens are also often associated with concepts of weak sustainability or light-green policies intended to ensure environmental protection within existing social, economic and political structures (Beckerman, 1994: 191–209). Blue-green thought has a variety of historical antecedents in the writings of Bacon, Malthus and Darwin and draws heavily on the writings of Adam Smith and John Stuart Mill (Smith, 1776; Mill, 1863). It is a reformist philosophy, since blue-greens suggest that environmental care is possible within existing structures. This reflects a commitment to conservatism that leads to preservation of the status quo which is easily translated into ideas about the natural balance in society, economics and politics. In addition, blue-green methods of environmental care are drawn from utilitarianism, liberalism and free market principles. The utilitarian principle of maximising the greatest good for the greatest number has been easily translated into methods of environmental management (Mill, 1863; Pearce et al., 1990; Winpenny, 1991). The importance of Enlightenment and Baconian principles wedded to conservative ideas has also led to a faith in technocentric environmental management. Technocentrism can be defined as the belief that the application of science and technology through planning or private enterprise is sufficient to control the environment. Technocentric planning to cope with environmental change has also been a significant feature of ecological management in the North and South (Drinkwater, 1991: 268–90; Sarre, 1995; Smith, 1996) and appeals to the ability to apply science in the realm of the environment are readily used to depoliticise the highly political issue of environmental conservation.

Second, red-greens can be loosely defined as those who are on the left of the environmental debate. Eckersley suggests that green political aspirations can be described as more left than right (Eckersley, 1992: 120). They share a common commitment with the blue-greens to sustainable development, but their analysis of why environmental degradation takes place and what policies are required to implement sustainable development is very different. Red-greens also include diverse groups, such as ecofeminists, who stress the kinship between women and nature, noting that women are more able to care for the environment because, in some way, they are closer to it (Eckersley, 1992: 63–71; Sen, 1995; McIntosh, 1996). Bookchin argues that when the environmental movement advocates protecting the planet from us, this masks real and important hierarchical social power relations and that the ecological impact of human reason, science and technology depends enormously on the type of society in which these forces are shaped and employed (Bookchin et al., 1991: 32). So for red-greens, it is social organisation which is the most important factor in explaining the state of the environment (Gorz, 1980: 20–8; Dryzek and Lester, 1989: 314–30).

Finally, the deep-green or deep ecology position claims to be a new third way, apart from socialist and capitalist conceptions of society, politics, economics and the environment. Deep-greens have been associated with notions of strong sustainability, which involve a radical break with existing social, economic and political structures in order to give environmental protection the highest priority (Beckerman, 1994). The work of James Lovelock on the Gaia hypothesis has been very influential in the deep-green debate. Lovelock suggests that the whole planet, including the mountains, seas and atmosphere, is a living organism (Gaia), that it is a self-regulating, self-organising system and that it is the greatest manifestation of life (Lovelock, 1979, 1988: 18–19; Leopold, 1986: 73–82). The animal liberation movement developed independently from the green movement but is now integral to it. In many ways the ideology of animal liberation is derived from the writings of Jeremy Bentham on the extension of moral obligations to all beings capable of experiencing pleasure or pain. Animal liberationists argue that all sentient beings are capable of experiencing pleasure and pain, and this leads on to the argument that animals should not be killed or exploited for human benefit (Singer, 1986; Eckersley, 1992: 33–49; Sterba, 1994).

But what relevance do these environmental ideologies have to the practice of wildlife conservation in Africa? The issue of wildlife conservation in Zimbabwe has been heavily influenced by differing environmental ideologies, and one important part of the development debate in sub-Saharan Africa is over the role and future of wildlife. Wildlife constitutes an important natural resource in southern Africa because it can provide revenue, meat and hides, and it forms the basis of the lucrative tourism industry in southern Africa. Wildlife also represents the wider question of how to ensure environmental management while remaining committed to development. Wildlife is central to the people-versus-nature debate in southern Africa and the importance of managing the relationship between animals and people is critical to successful environmental policy-making in Zimbabwe.

This book seeks to investigate the politics of wildlife conservation in Zimbabwe. First, it examines the relationship between ideas and policy formulation. In particular, the book seeks to demonstrate how the intellectual climate has shaped ideas of good conservation practice, and how these concepts, in turn, underpin the policy responses to specific environmental issues. Second, it analyses the way that the domestic and international contexts have influenced the formulation and implementation of conservation policies in the national context of Zimbabwe. In so doing, it highlights the competing political interests and the articulation of a political ideology for conservation. In sum, this book seeks to indicate the essentially political nature of conservation amidst rhetoric on an international level that presents it as an apolitical matter of saving animals.

1

Depoliticising & Repoliticising | Wildlife Conservation & Sustainability

The ideological underpinnings of environmental politics in the national and global arenas are made clear in differing conservation policy choices. In the case of Zimbabwe, the articulation of a policy of sustainable utilisation serves to negotiate a political settlement between potentially competing interest groups in the domestic context. A commitment to sustainable utilisation allows diverse interest groups, ranging from tourism businesses to local communities, state conservation agencies and local non-governmental organisations (NGOs), to be cemented together, because each group derives (albeit differential) benefits from the policy. Such appeals to ideas of sustainable use in Zimbabwe serve to depoliticise internal environmental politics and this then allows the local conservation movement to present a united front to potential critics within Zimbabwe and on the international stage by articulating a political and moral standpoint on sustainable utilisation, thereby actively repoliticising environmental politics to meet its own agenda.

Sustainable utilisation can be briefly defined as use of wildlife without jeopardising the continued survival of species. The policy of the Zimbabwe Department of National Parks and Wildlife Management (DNPWLM), hereafter referred to by its more popular and more memorable name 'the Parks Department', is that wildlife cannot survive in a developing economy unless it is economically self-supporting. In effect, wildlife use has become another form of land use, which is able to compete with crops and livestock. The policy objective is to induce commercial and subsistence farmers to view wildlife as an economic resource and, in turn, this will mean a greater area of land is made available for wildlife to allow wild populations to increase. In addition, in this way sustainable use aims to satisfy the demands for humane treatment of animals and concerns for human rights and development rights. However, the policy of sustainable use has a number of severe and powerful critics. First, this chapter will provide an analysis of the political context in which colonial policy was formulated. Secondly, it will examine the policy framework provided by ideas of sustainable util-isation in post-independence Zimbabwe. Finally, it will examine the crit-icisms of the theory and policy of utilisation, notably objections raised by notions of animal rights.

The policy and politics of wildlife utilisation

The political prominence of conservation policy in the post-inde-pendence period stems from the specific features of colonial and settler rule. The establishment of Southern Rhodesia as a commercial entity by the British South Africa Company (BSAC) from 1890 and the relatively early transfer of power in 1923 from imperial authorities to the white settler community had a significant impact on the political context in which policy was formulated. In the 1890s, reports by Lord Randolph Churchill and Hammond's Survey suggested that Rhodesia was not the gold-laden 'second Rand' that the Pioneer Column had hoped for (Rotberg, 1988: 335–6, 513–14). Consequently, agricultural interests were developed to compensate the white community, who had envisaged their involvement in lucrative gold mining. In order to protect settler agricul-tural interests, reserves (Tribal Trust Lands) were created for the indigenous populace to squeeze them out of the agricultural market. The reservation of productive agricultural land for white interests was crucial to the development of strong vested interests in commercial agriculture in Southern Rhodesia (Palmer, 1977; Rotberg, 1988). As a result, the question of land distribution came to be the central political factor in Rhodesia's history, which had a direct impact on the direction and form of conservation policy. For example, the land question was used to legit-imate interventionist conservation policies, such as draconian measures of cattle destocking in Tribal Trust Lands (Scoones, 1997: 621). Competition for land was a three-way process, incorporating powerful and powerless political interest groups, including the settler community, the indigenous populace and wildlife. The land question particularly affected wildlife conservation policy, precisely because wildlife required large areas of land, and so debates about conservation were visibly asso-ciated with the most controversial areas of colonial policy-making. Alongside the question of land, there was an extra constraint on the new government, namely the strong conservationist lobby in Rhodesian society, which had enlisted the assistance of international wildlife organ-isations to prevent the new government from deproclaiming national parks to provide land for resettlement (Stoneman and Cliffe, 1989: 180).

As a result of the racially unequal distribution of land in Southern Rhodesia, the divisions between settler and indigenous interests came to dominate numerous aspects of politics and policy-making. As with current conservation policy, a number of interest groups, including hunters, settler interests, central government interests, the growing tourism industry and international conservation organisations, were involved and each adhered to its own ideas of how best to manage wildlife. The settler and indigenous communities shared a common assumption that wildlife was a resource to be used. However, their views on what constituted legitimate utilisation of wildlife were vastly different. In general, the white community perceived wildlife as an object with recreational value – wildlife was to be hunted or viewed as part of one's leisure activities. What was known as the environment in the colonial context was constructed through varying environmental ideologies, including social Darwinism, the romantic imagery of hunting, conservationism, preservationism and technocentrism (MacKenzie, 1988; Neumann, 1996). The need to protect recreational facilities was in turn translated into a policy of excluding local people from safari areas and national parks. African knowledge and conservation techniques were ignored as 'backward' and 'ignorant', and undesirable environmental outcomes, such as degradation, were commonly attributed to them (see Leach and Mearns, 1996). The need for exclusion arose out of the view generally held by the indigenous community, that of wildlife as a source of food and materials. Alongside this view of wildlife as a resource, black communities also looked upon wildlife as a source of potential danger to crops, livestock and human life. These perceptions of the nature of the environment, coupled with the influence of settler agricultural interests served to provide a set of constricting principles that guided the direction of future wildlife policy-making.

The policy of sustainable utilisation is also partially a product of the particular political environment in the post-independence period. Policy had to be formulated within the constraints of a set of inherited institutions from the colonial period and within the context of promises made by the Zimbabwe African National Union (ZANU) during the liberation war of the 1970s. The land question was the central determining factor for environmental policy. During the liberation war, the guerrillas promised people in the Communal Lands that they would rectify uneven land distribution and make rural development a priority. However, in 1980, the Lancaster House agreement, which preceded independence, preserved the economic status of the settler community, and part of that was to ensure that white-owned land could not be expropriated by the state without compensation (Naldi, 1993; Government of Zimbabwe, 1994b: 29). The new government had to accommodate settler interests because they held a central place in the national economy. The post-independence government's policy of national reconciliation continued with concessions to the settler community, and promised structural change, such as large-scale resettlement, hence continued at a much slower rate (Stoneman and Cliffe, 1989: 37–51; Moyo, 1991).

Finally, during the transition from pre- to post-independence, the key personnel in the Parks Department did not change. This allowed for greater continuity in policy-making than might have been the case if the

Parks Department had been subject to the same pressures to immediately indigenise as other government institutions. The existing wildlife authorities were left intact by independence, even though a break with colonial practice might have been expected from a Marxist–Leninist post-independence government. In fact, the commercial objectives of wildlife conservation were retained from the colonial era. Drinkwater suggests that the legacy of the colonial period was more insidious than a simple black–white division of land and power. Rather, Zimbabwe's ruling élites were internally colonised by ideas of progress and development that arose from the colonial experience (Drinkwater, 1991). The fact that policy did not change significantly at independence is reflected in the carry-over into the post-independence period of many of the issues, problems and political conflicts that had arisen under colonial and settler rule. This history had a direct impact on policy formulation in the realm of the environment in post-Independence Zimbabwe.

The ideology of sustainable utilisation has been highly developed by Zimbabwe's wildlife policy-makers. The sustainable utilisation approach draws on the concept of a safe minimum standard of offtake of resources, and this ensures that the natural resource base is not depleted faster than it is replaced. It is closely related to the utilitarian idea that killing one or two animals will benefit the majority of wildlife. The starting-point for Zimbabwe's conservation policy is that natural resources exist in a social, economic and political context. Indeed, Zimbabwean state agencies and NGOs have sought to use the rhetoric of sustainable utilisation to bring political standpoints and moral questions to centre stage in the conservation debate. Notions of utilisation of wildlife carry a strong political and moral message, which defines animals as a resource to be used for human benefit. What the definitions clearly reveal is that the environment is conceptualised in terms of how it can enhance human welfare and is subordinate to human interests. This is in contrast to organisations and international treaties that tend to treat wildlife as though it existed in a social and political vacuum, and to notions of environmental science which imply that conservation does not affect people because it is purely about saving animals and habitats. Zimbabwe's definitions are related to red-green ideas and are diametrically opposed to deep-green and animal rights notions of preserving ecosystems and species for their own intrinsic value. Zimbabwe's policy focuses on the idea that natural resources are the basis of human existence and that they need to be conserved, but this does not mean that they cannot be used.

The ideology of sustainable utilisation is explicitly translated into policy documents and policy practice. The National Conservation Strategy (devised in 1982) sets out the definitions and parameters used by policy documents. Natural resources are defined as all forms of matter and energy available to the nation for development and the enjoyment of people and conservation is defined as the process whereby natural resources are managed in such a way as to ensure sustainable use and development, the latter being the process whereby the environment is modified to meet human needs and enhance the quality of life (Ministry of Natural Resources, 1990: 4). Natural resources are inextricably tied to the central issue of land distribution and land use: 'Land is the basis for all human survival and prosperity and civilisations have collapsed because they have not conserved it correctly' (Ministry of Natural Resources, 1990: 4). Conservation in terms of sustainable use is not intended to be in conflict with development concerns. Rather, they are mutually reinforcing processes since sustainable development directly depends on proper use of natural resources: 'Conservation is *not* in conflict with development but is the wisdom of sound and successful development' (Ministry of Natural Resources, 1990: 4 [their emphasis]). Sustainable use of natural resources is intended to remove the conflict between environmental care and economic development.

Wildlife utilisation is promoted by the Parks Department and the Ministry of Environment and Tourism as a means to redress the imbalances caused by colonial rule: 'Government intends to put into place mechanisms which ensure the equitable distribution of resources and access to opportunities for emergent entrepreneurship, without abrogating its fundamental obligation to society for conserving Africa's biological heritage' (Ministry of Environment and Tourism, 1992: 1).

Metcalfe argues that, as governments in developing countries have such difficulties in managing protected areas without the cooperation of local communities, wildlife policy should address questions of economic efficiency, environmental integrity, equity and social justice (Metcalfe, 1992a: 10). Sustainability requires an expressly political solution as well as the technical prescriptions of environmental science, and it is recognised that wildlife cannot be conserved in the face of antagonistic social and economic forces. Hence, the Parks Department is concerned that wildlife conservation is only likely to be successful if wildlife can be used profitably and the primary benefits accrue to people with wildlife on their land (Ministry of Environment and Tourism, 1992: 2). Wildlife utilisation is deemed to be the most effective means of establishing social and economic forces that are favourable to conservation (Ministry of Natural Resources, 1990: 9). The political settlement between potentially competing interest groups is achieved through the articulation of sustainable utilisation as a political ideology and (separately) as a means of defining a practical wildlife management policy. Zimbabwe's wildlife policy clearly has to accommodate a number of competing interests, including protected areas, rural poverty, land pressures and lack of finance (Metcalfe, 1992b: 4; Zimbabwe Trust, 1992: 1).

Zimbabwe's conservation policy is also legitimised by the Parks Department through reference to the favourable status of wildlife populations. If the numbers of wildlife are large, then more emphasis will be placed on its utilisation for human benefit and for the good of the environment as a whole. Utilisation is a means of ensuring that certain wildlife populations do not reach excessive numbers, which may result in damaging the environment or reducing biodiversity. For example, Zimbabwe's problem with elephants is not that the country has too few but, rather, too many. In the case of elephants, it can be argued that, since elephants are so numerous in Zimbabwe, their preservation value is very low (Barbier et al., 1990: 21). In contrast, because there are so few rhinos, preservation of the remaining ones is highly prized. In the Zimbabwean context, the government argues that utilisation of wildlife will affect population levels and composition, but it is not accepted that this is an argument against use *per se*. Wildlife is defined as a renewable (rather than a finite) natural resource. As a renewable resource, it can be used without destroying the resource base in the long term. Utilisation of wildlife relies on an offtake which is proportional to the size of a population, and economic returns that are reinvested into the species have to be beneficial for its status (Martin, 1994b: 3–8).

The Parks Department argues that the zero-use option, as represented by colonial legislation and preservation, manifestly failed across the continent. The reason, according to proponents of sustainable use, is that preservation was doomed to fail because it was based on an alien aesthetic framework and put colonial moral judgements about wildlife into a legislative framework (Zimbabwe Trust, 1992: 1). Preservationists and animal rights advocates adhere instead to the precautionary principle that wildlife should not be used, just in case its use threatens the continued survival of a particular species. Wildlife policy in Zimbabwe, however, is defined in terms of the context of aesthetic and ecological concerns, on the one hand, and economic, social and political practicalities, on the other (Zimbabwe Trust, 1992: 13; O'Riordan and Jordan, 1995). Wildlife policy-makers agree that the precautionary principle makes sound common sense when applied to global environmental issues, but that it may be disastrous when applied to ecosystems within African states where wildlife conservation has to compete with other forms of land use (Martin, 1994b: 11). If a sub-Saharan African government attempts to enforce a policy of zero use, utilisation takes place on an illegal and unsustainable basis (Martin, 1994a: 8). In addition, arguments in favour of preserving individual animals, rather than ecosystems and the numerous species they contain, can conflict with conservation objectives. Notions of animal rights have been criticised on the grounds that they are not in accordance with principles that ensure maximum biodiversity. For example, in allowing elephants to multiply, other important species, such as rhino, may be forced out of an ecosystem due to lack of food and water. As a result, Jon Hutton of the Zimbabwe-based Africa Resources Trust (ART) points out that it is important not to conflate individual animal welfare issues with conservation.[1]

Zimbabwe's policy also raises the issues of human rights and rights to development. Preservationist approaches imply foregone development

---

[1] Interview with Jon Hutton, Director of Projects, Africa Resources Trust, 13.3.95, Harare.

opportunities for land that is put under wildlife, and for industries involved in wildlife products. In Zimbabwe, the theory of sustainable utilisation is intended to ensure that wildlife contributes to development and to meeting basic human needs through use of wildlife revenues for community projects. It is a government priority that basic human needs and development rights are guaranteed before the rights of animals or the environment. The government sees this as a return to traditional precolonial rights to use wildlife in a controlled and sustainable way, particularly in the context of attempting to induce rural farmers in Communal Areas to conserve wildlife. The internationally regarded Communal Areas Management Programme for Indigenous Resources (Campfire) is deemed to work by the Parks Department and associated NGOs, because it harnesses and builds upon pre-existing African environmental ideas and traditions, rather than ignoring them (Kasere, 1995a). The articulation of a policy of sustainable utilisation in the Zimbabwean situation is very much related to human interactions with the environment. The lack of positive value that wildlife holds for rural people and commercial farmers is thought to lead to overexploitation of the resource. Similarly, conservation policy has to be socially sustainable and acceptable. Since wildlife entails so many social and economic costs for rural people, their approval for conservation policies is critical and without it, conservationists in Zimbabwe argue, wildlife is doomed (Metcalfe, 1992c: 1–7).

In terms of government conservation policy, the option chosen by the Parks Department is that of adaptive management (DNPWLM, 1992a: 2). At its most basic this can be defined as a trial and error process, where the policy is publicly legitimated on the national and international stages as management that is carried out according to practice and experience, rather than by adhering to models and theories. In addition, management policies are not executed on such a scale as to have irreversible effects, should the outcomes be negative (Zimbabwe Trust, 1992: 31–46). Adaptive management also rests on the principle that there are no truly natural ecosystems left, due to prolonged interaction with humans. Rowan Martin (former Deputy Director of the Parks Department) argues that there are very few truly natural or pristine ecosystems left on the planet and that most areas are inextricably affected by and linked with human activity. Consequently, for better or worse, human beings are involved in managing ecosystems, and sustainable utilisation of wildlife in Zimbabwe is an extension of that management (Martin, 1994a: 3–8). In practice, adaptive management has meant the increasing devolution of powers over wildlife and the strengthening of utilisation schemes by rural communities and ranches. In particular, Campfire is perceived to be the epitome of a pragmatic approach to the question of conservation and development. In fact, Campfire is intended to alleviate poverty and stress on the environment through use of wildlife (Wildlife Society of Zimbabwe, 1993). Tolerance of wildlife, and especially of elephants in rural areas, is increased by giving them an economic value, and hence elephant conservation is defined as an institutional issue as well as a technical one (Taylor, 1993: 14).

Rowan Martin has suggested that it would be wrong to begin with the assumption that Zimbabwe has the two policy options of preservation and sustainable use. He argues that use is unstoppable and the real question is not whether wildlife should or should not be utilised, but whether it can be turned into a useful resource.[2] For Parks authorities in Zimbabwe, it is clear that it is necessary for wildlife management policies to build on existing patterns of use, rather than attempting to halt or change them. As a result, Zimbabwe has made a number of policy choices that attempt to reconcile the demands of conservation and international responsibility for its wildlife heritage, with possibly conflicting demands for economic growth and social and political acceptability (Zimbabwe Trust, 1992: 2).

A number of states (including Botswana, Zambia and Mozambique) treat wildlife as something that is not owned by anyone. In contrast, with the 1975 Parks and Wildlife Act, Zimbabwe has taken a radical step by defining the landowner as the appropriate authority for managing wildlife, rather than the state.[3] The Parks Department is convinced that Zimbabwe's conservation policy has been successful because of 'a far sighted approach which recognises that landholders should be the best custodians of their natural resources provided they have the right to use wildlife and to benefit from their custodianship' (Ministry of Environment and Tourism, 1992: 4). Under the terms of the 1975 Parks and Wildlife Act, in the commercial farming sector the appropriate authority is the ranch owner, and in the rural context the District Council is the designated authority on behalf of the local community. This then allows the landowner to benefit financially, aesthetically, culturally and ecologically from wildlife conservation efforts in a direct manner. The 1975 Act also formally introduced the principle that economics could be used as a tool for turning wildlife into a resource. As a result, in the 1990s wildlife has outstripped cattle and crops in terms of economic value in many areas of Zimbabwe, but most importantly in ecologically fragile marginal lands (Jansen, 1990; Zimbabwe Trust, 1992: 31–46). Wildlife generates economic returns through its use as a tourist attraction, hunting trophies or a source of wildlife products, such as meat, skins, hides, horn and ivory. However, the Parks Department recognises that it is necessary to pay attention to animal welfare and thus the Department is not against any form of wildlife utilisation 'provided it falls within Zimbabwean Society's accepted norms of animal treatment' (Ministry of Environment and Tourism, 1992: 13). This is especially important, because the wildlife

[2] Interview with Rowan Martin, Assistant Director: Research, DNPWLM, 29.5.95, Harare.
[3] Parks and Wildlife Act 1975; and interview with Willie Nduku, Director, DNPWLM, 7.7.95, Harare.

authorities have been sensitive about charges of cruelty levelled against them by animal welfare organisations, because such adverse publicity has the potential to damage tourism and the sport hunting industry, which heavily rely on markets in Europe and America.

Political ideology and animal rights

Notions of animal rights have been readily mobilised by opponents of Zimbabwe's policy on sustainable utilisation. Ideas of animal rights are easily translated into snappy headlines and slogans, which effectively communicate an organisation's or individual's stance on utilisation. Consequently, it is important to understand the ideas that are used to legitimise anti-utilisation rhetoric in the international conservation movement. The diametrically opposed ethical viewpoints have led to heated debates on the international stage about conservation practice.

Opponents of Zimbabwe's conservation philosophy argue that it is essentially based on a model of nature that is hierarchical, because it accords human beings a primary position in that hierarchy, which gives humans dominion over wildlife. This is in conflict with ecological beliefs that the environment is a living system, of which humans are only one part. Opponents of utilisation argue from the standpoint that biodiversity is crucial. For present purposes, biodiversity can be defined as ensuring that ecosystems and all the species in them are conserved without allowing one species to dominate to the detriment of others (including humans). Value is calculated in terms of the role each component part plays in the ecosystem, rather than as an economic or use value. In accordance with this theory, all parts of an ecosystem have intrinsic value and so all parts have a right to be preserved (Lovelock, 1979).

Zimbabwe's allocation of a central position to human benefit and human welfare is fundamentally at odds with theories that place humans in a web of nature on an equal footing with wildlife and other environmental resources. In addition, this is related to the belief that wilderness areas are priceless and that with utilisation Zimbabwe is wrongly attempting to place a price on them (Sheldrick, cited in Care for the Wild, [1992?]). In effect, critics argue that wildlife has value beyond its economic worth, since environmental, aesthetic and cultural values are just as important. These ideas are extended to include the belief that human interference in the form of management of the environment upsets the natural balance. It is suggested that nature can correct its own imbalances and should be left to do so. Opponents of sustainable use argue that humans are not given stewardship over nature and do not have a right to interfere with it (Zimbabwe Trust, 1992).

The notion of a moral responsibility to preserve individual animals is also at odds with Zimbabwe's conservation policy. Animal rights organisations have redefined consumptive and non-consumptive uses of wildlife as non-lethal and lethal uses. Lethal uses of wildlife, such as sport hunting and culling, are opposed by animal rights theorists on the grounds that they are cruel and deny animals their intrinsic right to life. For some animal rights theorists, culling elephants is equivalent to culling humans. Animal liberationists argue that all sentient beings are capable of experiencing pleasure and pain, and so animals should not be killed or exploited for human benefit. Animal rights theorists suggest that policies and philosophies that value animals in terms of economics and human use will lead to conservation of selected useful species to the detriment of others (Eckersley, 1992: 33–49). Singer's concept of speciesism (that is, discrimination against other animals by humans) is often employed to argue that it is morally wrong to kill animals for human use or benefit. Singer suggests that, as intelligence and ability to communicate (verbal and non-verbal) does not entitle one human to exploit another, the same principle should be extended to animals (Singer, 1986: 24–32). These ideas are readily translated to elephants, which are capable of complex forms of communication. Animal liberationists point to elephants as an example of an animal that grieves the loss of family members and creates close bonds with other elephants in their group (Jordan, in Care for the Wild, [1992?]).

Conclusion

Zimbabwe's conservation policy is derived from a particular ideological basis that places the satisfaction of basic human needs above the rights of animals, and is closely related to ideas of people's stewardship over nature. Sustainable utilisation is also portrayed by its supporters as the only pragmatic solution to wildlife conservation in a developing economy in which wildlife must compete with other forms of land use in order to survive. The Parks Department argues that ultimately preservationism will lead to wildlife extinction and, moreover, that adherence to notions of animal rights, moral responsibility to preserve individual animals and non-interference in the balance of nature is in conflict with conservation of those species and their ecosystems. Instead, it is argued that conservation is as much about social, economic and political acceptability and practicability as about ecological or scientific principles. In order for wildlife to survive, people must accept it and acceptance will only come if wildlife has an economic value. In essence, the Parks Department's policy is that wildlife conservation results in development through the generation of revenues from sport hunting and wildlife-based tourism. This is in contrast with many other states, where wildlife preservation means foregone development opportunities for local communities, when financial and human resources and large areas of land are invested in wildlife, and tourist revenue is captured by the state or multinational capital, in the form of international tour companies.

Zimbabwe's policy of sustainable utilisation is presented by the government as a pragmatic answer to the twin concerns of conservation and development in a situation where there are many pressures on natural resources and government finance. However, the ideology of utilisation and its practice in Zimbabwe's wildlife management is heavily criticised from the animal rights perspective. The clash of these two ideologies of best conservation practice has a significant impact on the ways that Zimbabwe justifies its policy stance in the international arena, to potential donors, to international institutions and to international environmental NGOs. The debates over utilisation versus animal rights permeate the heated political clashes between the two camps over specific methods and examples of wildlife management, and this is what the book will focus on in the following chapters.

2

Individual Power & Personal Patronage | in Conservation Politics

A common complaint in the donor community and in wildlife charities is that the parks and wildlife departments in Africa are incapable of effectively conserving Africa's big mammals. Such ineffectiveness is often portrayed as a result of understaffing and poor financing in the face of onslaughts by greedy poachers. However, there are other factors that are internal to African wildlife agencies that have an impact on their performance. In Zimbabwe, important functions have been devolved to the private sector and to local communities, but the Parks Department has retained certain key roles, including regulating wildlife utilisation in the private sector and in local communities, and dealing with domestic security and the policing of wildlife laws. In addition, the Department manages relations with external donors and non-governmental organisations (NGOs), and it represents Zimbabwe's case in international fora, such as the Convention on International Trade in Endangered Species (CITES). The Parks Department is primarily responsible for managing the Parks and Wildlife Estate, and this includes wildlife conservation and population management, regulating tourism development and dealing with vegetation and water-supplies.

The Parks Department's capacity to carry out its management role has been influenced by a number of factors that are internal and external to the organisation. State agencies can be arenas for competition that is related to wider political and ideological divisions in a society. Conservation agencies prefer to present themselves as technical departments that are immune from political battles, but in Zimbabwe's case the Parks Department has provided a focus for wider political divisions that indicate a crisis of legitimacy for the ruling party. This chapter will first assess the Parks Department's role as a management organisation, including an analysis of its role in the Parks and Wildlife Estate. It will also examine the problems of funding and the process of commercialisation as factors that influence the Department's management capabilities. In the second part, this chapter will investigate the Parks

Department as a political organisation, and in particular its experience of power struggles.

The role of the state in wildlife conservation

For present purposes, the state can be defined as sovereignty and control over a continuous territory with clear boundaries. A state is also a political entity within which the citizenry is organised, with a set of interconnecting institutions and a bureaucracy that carries out state policy. Migdal characterises a strong state as one that has the ability to deliver economic development and a strong bureaucracy. He argues that the state in developing countries is weak, because other social groupings (such as landowners) are too powerful and constitute an alternative power base (Migdal, 1988). The African state has also been criticised for over-centralisation, due to being based on the statist model so fashionable in the 1950s and 1960s and a hangover from the colonial era (Wunch and Olowu, 1990).

Effective institutional arrangements are essential for the success of environmental planning, but government ministries are not necessarily organised in a manner that is suited to effective environmental management. For example, in Zimbabwe, there are separate ministries for Economic Planning and Development, for Lands, Agriculture and Rural Development and for Environment and Tourism. Environmental problems cut across the separated responsibilities of these ministries and require intersectoral planning and management. Ministries are organised in this vertical fashion because they concentrate on single policy areas, whereas environmental problems require horizontal management institutions precisely because they are affected by economic, agricultural and wildlife policies. Vertical organisations are less able to provide the level of coordination and cooperation demanded by environmental problems (Bartlemus, 1986). In addition, separate ministries may be unwilling to work together for all sorts of reasons, one of which is if they perceive the problem as one that belongs to their ministry. Another difficulty is that the recently established environmental ministries were accorded a rather junior position in government and, in Zimbabwe, the Parks Department is under the broad supervision of the Ministry of Environment and Tourism, although the level of interaction between the Ministry and the Department has varied with different ministers.

The Parks Department is the agency that is primarily responsible for managing wildlife resources and the land they live on. It manages the Parks and Wildlife Estate, covering 12.7 per cent of Zimbabwe's land area (Zimbabwe Trust, 1992). The Department's stated policy is to protect those wild resources and permit controlled utilisation for human benefit. However, although Zimbabwe's policy philosophy is sustainable utilisation, some preservationist policy areas remain. National parks are the most preservationist approach to wildlife conservation. Under the terms of the 1975 Parks and Wildlife Act, national parks are intended to preserve and protect the natural landscape and scenery, and also to preserve and protect wildlife and plants and the natural ecological stability of wildlife and plant communities in the park. The purpose of parks is to preserve habitats and species in a 'natural state'. In accordance with this, under the terms of the 1975 Parks and Wildlife Act, the definition of national parks enshrines the principle of inalienability: that is, parks represent a long-term commitment to conservation and they cannot be deproclaimed without a legislative change. The desire to conserve wildlife, especially large mammals, is apparent in national parks. Conserving flagship species, such as rhino, elephant and lion, is the *raison d'être* of many parks and reserves. Preservationist national parks have continued because they were set up during the colonial period, they are favoured by external donors and NGOs and, through tourism development, they are an important source of revenue for the state. National parks require substantial investment in terms of resources in the form of land, finance, human resources and law enforcement and they are favoured by governments and international conservation organisations, because they are a clear expression of a commitment to wildlife conservation.

The Parks Department's management role extends beyond simply ensuring that national parks exist. The Department has been involved in regulating tourist development, ensuring the integrity of park boundaries and protecting wildlife from poachers and, lastly, negotiating competing conservation and development interests within wildlife areas; it has also retained a regulatory role with regard to private tour companies operating in national parks. However, the Department is also part of the tourism industry because it owns accommodation and runs tours and hunts in national parks and safari areas. Tourism is an important factor in the maintenance of national parks in sub-Saharan Africa. The tourism industry in Zimbabwe is dependent on them, as visitors come to see wildlife, for which parks provide the major habitats. This revenue-generating capacity mitigates some of the very high costs of protection and maintenance of parks (Zimbabwe Trust, 1992). For example, in 1990 the World Tourism Organisation (WTO) estimated that tourism was worth US$62.5 billion for developing states, arguing that many of them were well placed to take advantage of ecotourism, which is the fastest growing tourism sector (World Tourism Organisation, et al., 1992: 1).

The Parks Department is responsible for ensuring the integrity of conservation areas. The principle of exclusion, which is integral to most national parks, is one of the most contentious aspects. One of the most striking features of national parks in sub-Saharan Africa is the commitment to total exclusion of people from within their boundaries (Prins, 1987). The 1933 London Convention reinforced this principle of total exclusion of local people (Olindo, 1974: 52–60; MacKenzie, 1988,

1990b). This grew out of the belief that the state can preserve habitats in a pristine condition and that rural people were not capable of managing the environment. The first concern demonstrates the principle that states have been considered the best environmental managers and that the state is willing to use coercion and military-style tactics to preserve ecosystems in a wild and human-free state (Peluso, 1993: 200). The second concern stemmed from colonial notions that it was the role of colonial administrators to teach local people about wildlife and conservation. Ironically, it is the exclusion of local people that has been one of the greatest threats to national parks. In the process of creating Zimbabwe's national parks, the Rhodesian government and the post-independence government actively removed people living in areas to be designated as national parks; for instance, when Gonarezhou was proclaimed a national park, the people living there were forcibly removed (Ranger, 1989; Moyo, 1991: 59). This stands in stark contrast to some parks in the North, such as the Lake District in Britain, which is a working landscape filled with people. This commitment to exclusion is reflected in all standard definitions of national parks. For example, the WTO defines national parks as 'a place not materially altered by human exploitation and occupation, and where the highest authority of the country has taken steps to prevent such exploitation except to allow visitors (tourists) to enter under certain circumstances' (World Tourism Organisation, et al., 1992: 2). In essence, national parks are intended to serve the potentially competing demands of allowing certain people access to the area while excluding other groups (Kenworthy, 1987).

The principle of exclusion has led to management problems. In many sub-Saharan African states, the boundaries of national parks have been completely breached by local people and commercial poachers. This has resulted in 'paper parks', which plainly exist on maps but in reality do not perform their primary function of protection of the wildlife and habitats they contain. The source of the antagonism is the legislation rooted in the colonial era, which removed access rights to areas that local people had utilised for years. Consequently, local people do not benefit from the traditional national parks system. Rather, it has meant that local communities have been denied access to important resources, while they still have to live with the costs of having wildlife so close to their crops and livestock. National parks not only indicate a vast investment by the national government, but also an investment by local people, who have to forego future opportunities to develop the area for their own crops and livestock (Zimbabwe Trust, undated a: 7). In Zimbabwe, as in the rest of the continent, the state has used paramilitary-style patrols in an attempt to prevent traditional hunting for food and gathering of foods and medicinal plants. Under the terms of the 1975 Parks and Wildlife Act, picking plants, hunting or destroying any wildlife or nests and selling any wildlife or plant specimens are forbidden, and the Minister of Environment and Tourism is required to 'do all things necessary to ensure the security of a park'. Indeed, in Hwange National Park, hunting impala and buffalo by locals for food has been a major source of antagonism between the park authorities and the people in the surrounding areas.[1] It is clear that the continued commitment to the traditional national parks system is out of step with the overall national conservation policy of sustainable utilisation of wildlife. However, to deproclaim national parks is not an option, because they generate foreign exchange through tourism, and it would be a deeply unpopular decision amongst donors, NGOs and probably large sections of Zimbabwean society. Instead, the large areas set aside in national parks are justified in terms of the broader policy of sustainable utilisation. They generate revenue through tourism, and are critical to the survival of wildlife utilisation, without which Zimbabwe would no longer be able to argue that the policy was a workable means of negotiating conflicts between parks and people.

In order to manage the Parks and Wildlife Estate, the Parks Department requires a substantial budget to carry out its duties effectively, and access to adequate funds is a critical factor that influences its capability. From its establishment, the Parks Department encountered problems with obtaining the necessary finance from central government. In the late 1980s and the early 1990s the financial crisis in the Parks Department deepened considerably. This was related to a number of national political factors and specific problems between the Department, the central government and potential private funders (NGOs and the World Bank).

Since 1990 Zimbabwe has been undergoing a World Bank-inspired structural adjustment programme, similar to those found elsewhere in sub-Saharan Africa. The Economic Structural Adjustment Programme (ESAP) was intended to result in a shift from a command economy to a market economy through various economic changes, including lowering external debt, reduced public sector spending and increased agricultural and industrial exports, liberalisation of foreign currency controls, abolition of price controls and deregulation of labour laws (Moyo, 1992: 321–9; Lopes, 1996: 19–23). The political leaders of Zimbabwe insisted that it was a home-grown version of structural adjustment, and so was tailored to the specific conditions of the country in order to cushion the effects of shock treatment used elsewhere. The national programme of public-sector cuts had a direct effect on conservation and the Parks Department. First, ESAP required cuts and pay freezes in all public-sector services. Secondly, overall budget cuts by the central government fell disproportionately on the Parks Department. The overall cut in government expenditure in 1993 was 6.4 per cent, but the Ministry of Environment and Tourism experienced a heavier toll, with a 15.5 per cent budget cut.[2]

---

[1] Pers. comm. UTC Operator, Hwange National Park, 23.6.95.
[2] *Financial Gazette* (Zimbabwe) 9.9.93 'Please Sir, May I Have Some More?'; *Africa Confidential*, 1997, 38: 10; Moyo (1992: 321–9).

The consequences of ESAP have meant that the public sector has had difficulty in retaining well-qualified staff, but the difficulties in the Parks Department definitely predated ESAP. During the 1980s, conservationists were concerned with the apparent loss of senior and well qualified Parks personnel to the private sector and overseas posts. From 1986 to 1988, the Parks Department was shaken by a series of resignations. The brain drain seriously damaged the Parks Department, since some of the most qualified conservationists in Zimbabwe were now working outside the Department rather than inside it. A series of explanations were offered for the series of damaging retirements and resignations. A number of those who left suggested that the poor salaries offered by government departments in relation to the private sector were to blame. It is clear that a number of ex-parks employees were attracted to NGOs and private companies for this reason. However, low morale and deteriorating conditions of employment were also highlighted. ESAP required further cuts in the civil service, which directly affected the Parks Department, which lost 250 staff as a result of a 1992 ESAP directive requiring all ministries to cut their expenditure by 10 per cent (Cunliffe, 1994: 1–11). The loss of personnel also meant that the Parks Department lost an important skills base, which affected the quality of training for new recruits and left a very significant shortfall of trained personnel for rhino protection and anti-poaching.

However, it is easy to lay too much blame on external financial institutions for public sector cuts in developing countries. In fact, the way in which public-sector cuts are allocated tends to be determined by national governments. In Zimbabwe, the central treasury and the central government decided where the cuts would fall. From 1988, annual budgets suffered dramatic declines: in national parks there was an operational expenditure (per square kilometre) drop from US$24 in 1988 to US$2.63 in 1993 (Dublin et al., 1995: 14). These figures do not provide a very clear picture of the magnitude of cuts that the Parks Department has faced in relation to overall government cuts. The decline in the budget occurred at the same time as the Zimbabwe dollar was devalued and inflation was rising, as a result of which the Parks Department had less and less revenue to pay for goods and services, which cost more and more.

Financial difficulties in the Parks Department may have been affected by internal problems, such as poor cost recovery, inefficient structures, poor resourcing and internal conflicts. The financial crisis in the Parks Department was also in part due to the failure of central government to allow the Department to retain some of the revenue it generates. Tourism was Zimbabwe's third largest earner before the drought in 1994–5, after mining and then agricultural products, and in 1995 it was estimated to be the second largest, knocking agricultural products into third place (Government of Zimbabwe, 1994a: 2). It was reported in the local press that Hwange National Park alone earns the country Z$400 million per annum.[3] What is clear is that the growth of the tourist industry has exceeded all previous estimates and is continuing to grow rapidly, but only a fraction of the revenue generated by tourism in the Parks and Wildlife Estate is ever returned to the Parks Department.

In order to improve the financial situation and to make the Parks Department a more efficient manager, the World Bank drew up proposals for restructuring. The World Bank suggested that the Parks Department should be decentralised, commercialised and allowed to retain more of its revenue from national parks. In 1993 it was reported that the government was seriously considering transforming the Parks Department into a parastatal, or even privatising it. The proposals for restructuring were part of an ongoing arrangement with the World Bank, which offered US$50 million for restructuring, for the Department to set up its own fund and become an autonomous body. At first, the Parks Department was to be privatised but this suggestion met with fierce crit-icism, so the idea of a statutory fund was developed. It was envisaged that the funds the Department was to retain could be used to improve services for tourists and improve anti-poaching and protection of black rhino, and the Department would be able to make more autonomous decisions.[4] The World Bank also donated Z$1 million to carry out feasibility studies on commercialising parts of the Parks and Wildlife Estate and tendering out services. In line with other plans to restructure the public sector and the economy, the Parks Department was targeted to make its operations more efficient, improve management and reduce costs.

The proposals to commercialise the Department were highly contro-versial and became stuck in central government. Caesar Chidawanyika of the World Bank commented that internal restructuring of the Department had proved to be the stumbling block for commercialising national parks.[5] The controversy arose out of the involvement of two powerful sets of interest groups. Supporters suggested that restructuring would wrest control of the Parks Department out of the vagaries of central government, allow it to retain the revenue it generates and free the Department from corruption. Conservationists in Zimbabwe, while concerned that wildlife should not be exclusively in private hands, were aware that there was a need for change in some form in the Department.[6]

---

[3] *Financial Gazette* 2.2.95 'Obduracy Threatens Natural Heritage'.

[4] Interview with Caesar Chidawanyika, Senior Programme Officer, World Bank, 3.5.96, Harare; interview with Stephen T. Bracken, Second Secretary, US Embassy, 21.3.95, Harare; see also *Herald* (Zimbabwe) 21.3.95 'Fund For Parks in the Offing'; *Herald* 23.3.95 'Minister Squashes Rumours of Privatising National Parks'; *Herald* 8.4.94 'National Parks Department Seeks Autonomy – Minister'; *Sunday Mail* 16.1.94 'Funds Sought for National Parks'; *Financial Gazette* 19.8.93 'National Parks to Go Commercial?'; *Financial Gazette* 28.4.94 'Proposals to Boost Anti Poaching Units'.

[5] Interview with Caesar Chidwanyika, Senior Programme Officer, World Bank, 3.5.96, Harare.

[6] *Herald* 23.3.94 'Time for Action on Corruption'; *Financial Gazette* 24.3.94 'Department Replete with Corruption'.

Supporters of commercialisation argued that, unless the Department was commercialised, the government would retain control until the parks were unable to function and conservation efforts broke down completely, due to financial strain.[7]

The moves to commercialise and decentralise management in the Parks Department were strongly resisted by a powerful interest group. The individual personalities involved in these battles are not the main issue, since they formed part of the shifting membership of sets of alliances that remain the most important structures for conservation in Zimbabwe. However, the value of an analysis incorporating the personalities involved does provide a discussion of the personal networks that underscore environmental policy-making, a process that is often ignored in the literature about the environment, because of the tendency to attempt to remove it or raise it from the realm of politics. Those who argued against the proposals formed an influential alliance, which included the then Permanent Secretary to the Ministry of Environment and Tourism, Tichafa Mundangepfupfu, the Deputy Director of Parks, Willis Makombe, and their allies in the Zimbabwe African National Union – Patriotic Front (ZANU-PF). A number of other senior Parks officials were opposed to decentralisation and commercialisation, because they feared a loss of control over Parks Department decisions and, most importantly, resources. Commercialisation was perceived as privatisation by stealth and as intended to prevent certain interest groups from obtaining control over lucrative natural resources and tourism developments. Mundangepfupfu was openly opposed to restructuring and argued that Parks should stay in public hands, motives for opposing commercialisation which have been questioned. Supporters of commercialisation, such as Deputy Director of Parks, Rowan Martin, and Parks Director, Willie Nduku, argued that the proposals were blocked because they threatened his empire. Mundangepfupfu had one of the longest reigns in the public sector, being Permanent Secretary from 1987 to 1996. A number of senior Parks officers complained of his attempts to retain influence through interference in the day-to-day running of the Department. Mundangepfupfu was closely associated with the Deputy Director of the Parks Department, Willis Makombe, and also with Joshua Nkomo in the central government.[8] Those in Parks who opposed the restructuring were backed by more powerful elements in other ministries and in the Cabinet. It was reported that the proposals were directly blocked by the Cabinet Development Committee, which is comprised of powerful bureaucrats backed by Joshua Nkomo.[9] The Ministry of Finance was also reported to be attempting to block the restructuring of the Parks

---

[7] *Financial Gazette* 2.2.95 'Obduracy Threatens Natural Heritage'.

[8] Anonymous interviewees.

[9] *Financial Gazette* 24.11.94 'Wildlife Proposals Derailed'; *Herald* 21.3.95 'Fund For Parks in the Offing'.

Department because the Parks and Wildlife Estate is a major income earner for the treasury, and any move to change this would result in reduced revenues for it.[10]

The proposals were not approved until 1995 and the supporters of commercialisation explained this as the result of the influence of their opponents in central government. The proposals, they alleged, were swallowed by bureaucracy and sabotaged by internal ZANU-PF politics. The World Bank proposals had to be approved by the Cabinet in order to be implemented, but they languished at Cabinet level from 1993 to 1995. David Cook, the World Bank representative in Zimbabwe, stated that the restructuring proposals were slowed down due to problems of differing views in the relevant ministries and in the Parks Department itself.[11] It is significant that the proposals were finally approved in 1995, once Rowan Martin and Willie Nduku were suspended and Deputy Director George Pangeti had left the Parks Department. Caesar Chidawanyika commented that the suspension of Nduku and Martin had a definite impact on the momentum for restructuring, since the plans were approved by Cabinet just two weeks after the suspensions.[12] This was perceived as a political trade off between differing sets of interest groups. The price of approving the restructuring proposals was the removal of the most vocal critics of ruling-party influence on conservation decision-making. Clearly, with the amount of finance that is involved, there will be tensions over how that revenue is spent, but these conflicts over the restructuring were part of a bigger power struggle, which rapidly turned the Parks Department into a political liability on the international stage of conservation politics.

The Parks power struggle and patronage alliances

The power struggle within the Parks Department was indicative of wider divisions in Zimbabwean politics. It illustrated the lines of patronage in the ruling party, attempts to gain control over resources by certain alliances and the use of racial divisions as a cover for corruption. Bayart suggests that surviving precolonial social systems appropriated colonial institutions at independence, and that it is this appropriation that gave the present African state its historicity (Drinkwater, 1991; Bayart, 1993). Precolonial social structures, such as lineage and ethnicity, continue to exert an influence on the state today. Hyden argues that the African state is bottom-heavy, because it is a rhizome rather than a root system (Hyden, 1990). Makumbe argues that the colonial legacies in African countries

---

[10] Interview with Stephen T. Bracken, Second Secretary, US Embassy, 21.3.95, Harare.
[11] *Financial Gazette* 24.11.94 'Wildlife Proposals Derailed'.
[12] Interview with Caesar Chidawanyika, Senior Programme Officer, World Bank, 3.5.96, Harare; see also *Financial Gazette* 17.8.95 'Plans to Restructure Parks in Progress'.

play a major part in determining the role of the state in conflict resolution among elements of class, culture and social and economic interests. As a result of a weak state bequeathed by colonialism, in most African countries the state has been privatised or personalised, and its role in being an impartial arbiter in cases of abuse and misuse of public resources is limited (Makumbe, 1994: 59). The size of the patron–client network is affected by the volume of favours (or prebends) available. Clientelism often intersects with ethnic identity (Lemarchand, 1988), affecting all areas of the policy process and influencing the nature of environmental policy, because natural resources and licences to exploit environmental goods can be used as personal favours to reward key elements in the networks of patronage that permeate nominally public institutions.

Taken at face value, planning makes a great deal of sense to state agencies, but what happens in reality is often very different. Bureaucrats have the power to allocate rights over scarce resources and it is unsurprising that lines of patronage spring up under such circumstances. However, the political context in which these networks are established is very important, as the balance of power within a state institution and in a society directly determines what is demanded and by whom (Ades and di Tella, 1996: 6–11; Khan, 1996: 12–21; Coolidge and Rose-Ackerman, 1997). Policy élites can shape debates and decisions on a particular issue. There is clearly a discrepancy between what is planned and what actually happens in Zimbabwe, and this policy slippage is characterised by an implementation gap between what the people at the top (senior ministers) think is going on and what the people at the bottom receive in a conservation project. Policy slippage has been a focus of donor concern, where loans or aid is given for projects and new policies. The quality of African management has often caused outrage amongst donors, who wish to see concrete results for the loans and grants they offer to improve environmental care. Policy analysts and the donor community erroneously view the process of policy slippage as something irrational, when in fact it is perfectly understandable. This is what Bayart termed 'the politics of the belly', meaning that policy élites will not act against their own interests, and recognising that the use of prebends to support patron–client networks is an effective means of ensuring political survival (Bayart, 1993).

Resources are managed in a way that reflects these lines of patronage. Clientelism has a clear role to play in Zimbabwean politics and especially in the case of the Parks Department power struggle. The power struggle relied on rumour and allegations in the media. Ellis argues that rumour has been an important force in African politics, partly because, in the absence of a free and critical media, conversation with friends is often the source of the newest political news. In African states, information that is passed through spoken networks is more trusted than in Western societies, which cling to the idea that real news is found in newsprint (Ellis, 1989). Unlike newspapers, radio and television, rumour cannot be banned, closed down or manipulated to exclusively serve the interests of the state, but instead it can provide one of the few unregulated forms of political communication in societies where the media is strictly controlled. The following account of political conflicts within the Zimbabwean Parks Department inevitably draws on rumours that are not independently verifiable. However, the role of rumour as an instrument of political communication is significant in itself, and illustrates the conflicts that have arisen. The two factions have used the media and other channels to spread rumours and allegations to undermine each other. The fact that the rumours were believed by NGOs and donors demonstrated the importance of political accusations in Zimbabwean politics.

The two groups involved in the power struggle in the Department can be characterised as the 'conservationist alliance' and the 'patronage alliance'. The conservationist alliance included the former Minister of Environment and Tourism, Herbert Murerwa, senior Parks officials, including Rowan Martin and Willie Nduku, and their supporters within the Department. This group was highly respected in the international community and amongst local conservation NGOs. They are all qualified and skilled professionals and have had a long record of commitment to wildlife conservation in Zimbabwe. Their domestic and international standing meant that they were able to attract donor funding for wildlife programmes and represent Zimbabwe's case at CITES.

The opposition to the conservationist faction formed an influential alliance, which included the Permanent Secretary to the Ministry of Environment and Tourism, Tichafa Mundangepfupfu, Willis Makombe in the Parks Department and their allies in ZANU-PF. Graham Nott, the head of the Investigations Branch in the Parks Department, has also been implicated, in the process of pursuing his own agenda, which is unclear, but in so doing he reinforced the position of Willis Makombe against Rowan Martin and Willie Nduku.[13] The ability of this alliance to exercise control over the Parks Department was derived from its domestic political position. Wildlife conservation is one of the most racially controversial areas of public policy in Zimbabwe, because conservation has a tradition of being perceived as a white domain and its demands for large areas of land means it is in direct conflict with the social and political aspirations for land redistribution. Superficially, the patronage faction used the issue of indigenisation and black empowerment to justify its actions against the conservationist faction, which was largely identified (though not exclusively) with the white settler community. This proved particularly effective in the run-up to the 1995 and 1996 elections. Its influence was also derived from the support it received from the Vice President, Joshua Nkomo. Nkomo occupied a special position in Zimbabwean politics, which allowed him to engage in a

---

[13] Anonymous interviewee.

number of activities without question or investigation. The 1987 Unity Accord brought ZANU-PF and the Patriotic Front-Zimbabwe African People's Union (PF-ZAPU) together after five years of civil strife, and this meant that former PF-ZAPU leaders were incorporated into the government and state structures. Nkomo was leader of PF-ZAPU, which represented a largely Ndebele constituency. Since Mugabe was unable to gain control over Ndebele areas, he invited Nkomo to be Vice President and PF-ZAPU to join the ruling party in an alliance (Stoneman and Cliffe, 1989; Moyo, 1992: 314–15; Makumbe, 1994: 59–60). Joshua Nkomo was involved in wildlife policy through the Cabinet Development Committee and the Development Trust of Zimbabwe, two bodies involved in attempts to obtain certain parts of the Parks and Wildlife Estate for personal gain.

In the course of 1995, the power struggle in the Parks Department became extremely intense and highly public. The power struggle has continued, but its peak was in 1995, when the shifting alliances in the ruling party and in the Parks Department provide an example of how personalised patron–client relationships directly affect wildlife conservation. It manifested itself in a number of ways that highlighted the wider issues faced by the public sector and by conservationists. The power struggle was cloaked in debates about indigenisation and black empowerment. Conservation is perceived as a racially charged black–white issue by both black and white Zimbabweans. In the transition to independence, one of the departments left almost intact was the Parks Department. One commentator suggested that it was unusual for a black government to leave such a level of decision-making almost entirely in white hands, the reason being that one élite passed power to another without considering their previous commitment to restructure the government and economy to allow black Zimbabweans a fair chance.[14] Herbst (1990) suggests that, if an institution was created prior to independence, it is likely to be white-dominated and unlikely to be trusted by black Zimbabweans.

The power struggle had its roots in the series of resignations and retirements in the late 1980s. As a result of losing so many senior Parks officers, a vacuum was left, which was not filled by suitably qualified personnel. The recruitment drive and promotions that followed the brain drain acquired a racial overtone. In the post-independence period, the public sector has continually been undergoing a process of indigenisation, providing black employees the opportunity to obtain promotion and experience in order to redress the racial imbalance in employment caused by the settler regime (Lopes, 1996: 25–6). Indigenisation is highly controversial and the conservationist faction argued that, while addressing such an issue was perfectly legitimate, the indigenisation

---

[14] Pers. comm. Daniel Campagnon, Institut François des Recherches Africaines (IFRA), 28.2.95, Harare.

debate had been purposely subverted and was used as a screen to hide clear cases of patronage and corruption, and that white Parks officers were squeezed out to make way for those allied to the elements in the ruling party under the guise of black empowerment. The process of the brain drain continued into the 1990s, and it seriously damaged the reputation of the Parks Department, because it was perceived that people who were not qualified had received rapid promotion. The appointment of Willis Makombe as one of the Deputy Directors was perceived by the conservationist faction as the result of indigenisation used wrongly to promote the unqualified, but politically favoured. The conservationist alliance raised concerns that without suitably qualified personnel the Department would not be able to carry out its most basic functions.[15]

Conversely, the privatisation and commercialisation proposals for the Parks Department were perceived by some elements in ZANU-PF as intended to empower the white community once again. There are very few prominent black figures involved in private conservancies. The growth of wildlife utilisation has resulted in white commercial farmers obtaining stocks of wildlife from the Parks and Wildlife Estate, so, when white officers have left the Department to enter the private sector, those interested in indigenisation have argued that these white officers have merely stocked private conservancies and game farms in their own interests and in the interests of their friends.[16] However, although it has taken on a racial theme, conservationists, donors and a number of Parks employees agreed that the debate was not really about racism in the Department or black empowerment. Indigenisation was used as a smoke-screen, behind which the patronage alliance could manoeuvre their way into positions of power. The power struggle was about attempts to control resources that the Parks Department is able to dispose of, and hence about access to political power within Zimbabwe.

The power struggle was a political struggle between rival groups in conservation and in ZANU-PF. One commentator suggested that Willie Nduku was harassed because his previous political affiliations should mean that he was not accorded such a high status in a government department. In the 1970s he was closely linked to Reverend Sithole, an outspoken critic of the Mugabe regime and head of a small opposition movement, ZANU-Ndonga.[17] The senior management of the Parks Department was not immune from investigation and harassment. For example, Deputy Director Rowan Martin was subject to three differing sets of corruption charges. The first charges related to his acceptance of a

---

[15] *Financial Gazette* 18.12.87 'Suspicion and Distrust Sadly the Order of the Day at Parks Department' (letter from 'depressed former ranger'); *Financial Gazette* 9.12.93 'National Parks No Longer Capable'.

[16] Interview with Tom Taylor, Chief Executive of the Save Valley Conservancy, 10.5.96, Harare. The issue of black empowerment in the private conservancies is dealt with more fully in the chapter on private wildlife interests.

[17] Pers. comm. from anonymous source.

vehicle in return for allowing Zimbabwean black rhino to be exported to
international breeding programmes in Australia and the USA. Nott
claimed that Martin had accepted a Toyota Land Cruiser from Ingrid
Schroeder of the International Rhino Foundation in order to smooth the
agreement in 1991, and so had bartered the rhino for a vehicle. However,
Graham Nott was barred from investigating Martin's case and Nduku
claimed that he only found out about Nott's investigations when he heard
of them from the USA. Martin denied he sold the rhinos but agreed that
he had misjudged the situation in accepting the vehicle.[18] The second set
of allegations, again relating to acceptance of vehicles, resulted in Martin
being sacked, for which he sued and the allegations were dismissed. The
third set of accusations of misappropriation of funds were the result of
elephant translocation. Willie Nduku and Rowan Martin were suspended
as a result of these accusations and, despite the fact that the claims
remained unproven, it is unlikely that either will return to the Parks
Department. The power struggle reached its height in 1995 after the
Director and Deputy Director of Parks, Nduku and Martin, were
suspended, resulting in resignations by professional conservationists
from within the Department. The suspensions were supposed to last for
six months while investigations were carried out. In fact, Rowan Martin
was reinstated in 1997 but was then moved to another post in the
Ministry of Environment and Tourism, and he resigned because he was
unhappy with such a change. The harassment of senior Parks officials
reached a peak in 1995 with the Gordon Putterill case.

Putterill was the former Chief Warden of the troubled Gonarezhou
National Park in the south-eastern lowveld of Zimbabwe. The Putterill case
became synonymous with the power struggle in the Parks Department, and
it also provides one of the clearest examples of the influence that personalised political struggles can have on effective implementation of conservation policy. It was clear that it was important to some interest groups that
he was removed from his post, since numerous disciplinary cases were
brought against him. For example, in 1992 he was reprimanded for shooting
100 cattle that had strayed into Gonarezhou in search of food during the
drought (although his actions were eventually found to be lawful) and he
was arrested and detained for shooting a poacher during routine anti-
poaching operations.[19] Those who supported Putterill perceived these cases
as attempts to remove him because he had proved an obstacle to the
patronage alliance, as well as to poaching interests in the area.

However, it was the elephant export scandal that proved to be the end
of the power struggle. The scandal directly removed Putterill, Martin and

[18] Interview with Stephen T. Bracken, Second Secretary, US Embassy, 21.3.95, Harare; see also
Herald 28.9.94 'Parks Director Accused of Protecting Deputy'; *Daily Gazette* 28.9.94 'Nduku
Backed Accord on Rhino Relocation'; *Daily Gazette* (Zimbabwe) 27.9.94 'Murerwa Denies
Being Lobbied'.
[19] *Herald* 16.7.92 'Game Warden Reprimanded'; *Herald* 2.9.92 'No Arrest Made in Gonarezhou
Shooting'.

Nduku. Indirectly, it meant the loss of other conservationists in the Department, who were unwilling to continue to work for the Parks Department, and it is significant that almost all of those interviewed in their capacity as professional conservationists in the Department left in the period between July 1995 and May 1996. In 1993, Putterill, Nduku and Martin were the subject of an investigation into their involvement in the elephant export scandal, which involved a claim by Graham Nott that the government could have lost Z$2.5 million in the sale of live elephants from translocation exercises in Gonarezhou National Park. In 1995, this case was brought before the Public Service Commission and was later referred to a parliamentary committee under Chris Kuruneri, MP.[20] The case pitted the two factions against each other, revealing deep divisions in the Parks Department and turning wildlife conservation, corruption and the state of the Department into a highly public debate.

These events arose during the 1992–3 drought, when Gonarezhou National Park was the centre of a massive elephant translocation exercise. Willie Nduku gave authority to translocate 500 elephants, some of which were due to be exported to South Africa. Nott was concerned that 206 elephants were sold at a price that was below the market value to a group of farmers, through the Chiredzi Branch of the Wildlife Producers Association. These farmers then sold the elephants on to a newly created park in Bophutatswana. Nott and his allies argued that Willie Nduku corruptly handed translocation contracts to the private company Wildlife Management Services, in breach of Tender Board regulations, and Nott alleged that the money for the exported elephants was paid into a Swiss bank account. However, Nduku rejected this, stating that all banking was done through a local bank in Chiredzi.[21] An inquiry was launched into the affair in 1994 to investigate whether the Parks Department and Wildlife Management Services had clandestinely exported elephants and expropriated the funds. The investigation was carried out by the Comptroller General and the Auditor General before the case was passed to the Public Service Commission. Suspicions were raised because Putterill was part of the Gonarezhou Drought Crisis Committee, which was responsible for overseeing the translocation, and also because he had worked closely with Wildlife Management Services, which was headed by a former Parks employee, Clem Coetsee. Finally, it was reported that Gordon Putterill was accused of misuse of public funds, with the complicity of Nduku and Martin, who had not sought the authority of the Ministry of Environment and Tourism for the elephant exports.[22] In 1995, as a result of the case, all translocations were halted except those deemed

---

[20] *Herald* 7.2.96 'Parks to Come Under Scrutiny'.
[21] *Financial Gazette* 23.12.93 'Elephant Export Scandal'.
[22] *Herald* 25.12.93 '*Murerwa Probes Elephant Exports*'; *Financial Gazette* 13.1.94 'Probe Into Elephant Deal Launched'; *Financial Gazette* 15.9.94 'Diplomatic Row Looms Over Fund'; *Sunday Gazette* (Zimbabwe) 16.1.94 'Poor Accounting Costs National Parks Millions'.

absolutely necessary to prevent ecological damage. Before their case was heard, Nduku and Martin were suspended from duty for six months under the orders of the new Minister of Environment and Tourism, Chen Chimutengwende (who by 1999 had become the Minister of Information).[23] They were not reinstated after the six month period, because the case was still being heard.

One of the main charges that Putterill faced was that he had failed to channel funds through the central treasury when he should have. The charge of defrauding the state of its rightful finance was seen by those who supported Putterill as trumped up. A number of donors and NGOs were involved in the translocation exercise in 1993. The German government provided some of the funding for translocation through their overseas development fund, Deutsche Gesellschaft fur Technische Zusammerarbeit (GTZ). Other funds were channelled through a number of NGOs, including the Zimbabwe Trust, Tusk and Care for the Wild. Once the investigation was launched, these organisations made their frustration with the situation in the Parks Department public. The German government accused the Zimbabwean government and the Investigations Branch of breaking the spirit of cooperation embodied in an agreement signed between the two states in 1981, while the GTZ was angry that funds donated by them and held in a local bank account were frozen on the orders of Chimutengwende and argued that such funds were not the property of the Zimbabwean government.[24] The US Fish and Wildlife Service also paid some of the translocation costs, and made it clear that the money was to be paid directly to Clem Coetsee of Wildlife Management Services by the Parks Department. This meant that none of these funds were ever intended to be passed through the central treasury (although in principle they should have been), because the donors believed that, if the treasury was involved, the funds for translocation would never actually reach their intended destination.[25]

Those who supported Nduku, Martin and Putterill raised doubts about the way in which the case was conducted. Putterill's case was a Public Service Tribunal, and so was not heard under normal court conditions.[26] However, after the Tribunal began, the case was passed to the police Criminal Investigations Branch which resulted in Nduku and Martin being called in for questioning, pending a court case.[27] Therefore, the case did not follow the normal course for criminal investigations, being closed to the public and the media. Putterill's allies also questioned the way evidence was collected against him. One of the chief witnesses for the prosecution was from the Parks Investigations Branch, which the

[23] *Herald* 11.5.95 'Ministry Bans Exportation of Wild Animals'.
[24] *Financial Gazette* 15.9.94 'Diplomatic Row Looms Over Fund'.
[25] Interview with Stephen T. Bracken, Second Secretary, US Embassy, 21.3.95, Harare.
[26] Herald 21.3.95 'Jumbo Sale Tribunal Resumes'.
[27] *Financial Gazette* 6.7.95 'Two National Parks Bosses Suspended'; and *Herald* 6.7.95 'Ministry Suspends Two Parks Bosses'.

conservationist faction perceived as allied to Willis Makombe and his supporters. One supporter of Putterill claimed that the chief witness tortured one of Putterill's scouts in Gonarezhou to make him say that Putterill was corrupt. As a result of this information, there was an unsuccessful appeal to the High Court about the case.

Those who supported the translocation exercise in the donor community and in the Parks Department argued that the charges against Nduku, Martin and Putterill were drawn up in order to discredit them and remove them. Willie Nduku commented that he felt the motivations behind his suspension and the allegations against him were unclear. He suggested that it was either genuinely about a need to ensure accountability in the use of public and private funds and services or it was motivated by elements in the Department that wished to have him and some of his colleagues removed. He agreed that the latter was the most likely explanation and he said that he believed neither he nor Rowan Martin had committed a crime in authorising the exports and the involvement of Wildlife Management Services.[28] Indeed, there was some evidence to suggest that the allegations were part of the power struggle. For example, the timing of Minister Chimutengwende's announcement to halt translocations until all cases of corruption in translocation and wildlife sales were investigated was significant. Chimutengwende was sworn in as the new Minister of Environment and Tourism when Nduku and Martin were out of the country dealing with CITES related business. The conservationist alliance argued that, in the meantime, the opponents of Martin and Nduku approached the Minister and persuaded him to halt the translocations and order an inquiry into the allegations of corruption in the elephant export scandal.[29] Rowan Martin commented that he thought that the halt to translocation would be lifted fairly rapidly, because the new Minister had been persuaded by a coalition of 'very nasty people' from the Department, other ministries and the private sector that the Department was in chaos and that he and Nduku were to blame.[30] It was clear that a number of key conservationists in Zimbabwe perceived the elephant export scandal to be another round in the power struggle between factions in the government. It was not insignificant that, following the suspensions of Martin and Nduku, Willis Makombe was left in charge of the Parks Department. The other Assistant Director, George Pangeti, was not chosen and eventually resigned from the Department.

[28] Interview with Willie Nduku, Director, Department of National Parks and Wildlife Management (DNPWLM), 7.7.95, Harare, who was suspended from 10 July 1995 for six months. Nduku was unable to comment in detail on the case because it was an ongoing criminal investigation.
[29] Interview with Rowan Martin, Assistant Director: Research, DNPWLM, 29.5.95, Harare; interview with Brian Child, head of the Campfire Coordination Unit, DNPWLM, 16.5.95, Harare.
[30] Interview with Rowan Martin, Assistant Director: Research, DNPWLM, 29.5.95, Harare.

The removal of Gordon Putterill as warden at Gonarezhou was also highly controversial. He was closely associated with Martin and Nduku and their supporters in the Parks Department and in NGOs. Putterill was also perceived by some nationalist elements in ZANU-PF to be selling the country's natural heritage off to white farmers. Minister Chimutengwende received congratulations from these elements for refusing to allow the heritage to be sold to white farmers when he banned wildlife sales and translocations in 1995.

However, those who have supported Putterill and those who have investigated poaching in Gonarezhou suggest that certain factions in the Parks Department and central government have another reason for his removal. Putterill had a long history of service in Gonarezhou National Park, being warden during the late 1980s when there were reports of official involvement in poaching in the area. There were persistent rumours of a poaching faction in the Department [31] and that this faction was keen to see Putterill discredited, so that any information from him was also discredited. As part of their campaign to discredit Putterill, his opponents claimed that he had created a personal empire in the national park and that all his rangers and scouts were frightened of him.[32] Putterill claimed that numerous attempts were made on his life and that he was victimised and harassed. However, some of his supporters have also pointed out that Putterill had an almost fanatical attachment to Gonarezhou and that his continual allegations of the authorities attempting to murder him verged on complete paranoia.[33]

The case and the highly public power struggle had a damaging effect on the image of the Department. Even if senior Parks officials were guilty of selling the elephants at a low price and awarding contracts without respecting Tender Board rules, the activities of those attempting to discredit them were far more damaging. For example, the donor community became disillusioned with the capabilities of the Parks Department and certain donors preferred to fund NGOs involved in wildlife rather than the Department. Jon Hutton argued that the Parks Department was a liability in the international arena, since the constant divisions and infighting damaged the level of influence that Zimbabwe's argument had for reopening a legal ivory and rhino horn trade at CITES.[34]

A second case, the VIP hunting scandal, illustrates the level of control that the patronage alliance has in the Parks Department and also illuminates the peculiar position that Joshua Nkomo held in Zimbabwean politics.

---

[31] Anonymous interviewee; see also *Daily Gazette* 11.10.93 'Top Officers Accused of Helping Poachers'.
[32] *Financial Gazette* 8.6.95 'New Twist to National Parks Power Struggle'.
[33] Anonymous interviewees.
[34] *Daily Gazette* 26.11.94 'Put House in Order, Zimbabwe Urged'.

Zimbabwe is very sensitive about its record on sport hunting. Hunting is unpopular as a conservation method in Western states and NGOs, and so the Parks Department has been keen to demonstrate that hunting is carried out with ethical and moral considerations in mind. For example, hunting female elephants is outlawed because the impact of shooting matriarchs would be damaging to elephant social systems and many genes for big tusks are stored in female elephants. Due to the international publicity surrounding elephants, Zimbabwe has tried to improve its image by stopping hunting females (Child, B., 1995d). The Parks Department has been keen to try to prevent unethical hunting practices, and media exposure of hunting practices in the hunting scandal had a damaging effect on its image and position as a major proponent of sport hunting for conservation purposes in the international arena. The conservationist alliance in the Parks Department believed that Joshua Nkomo was keen to exercise control over certain resources and areas held in the Parks and Wildlife Estate, especially around Lake Kariba, in order to develop a tourist centre in the area. The suggestion was that he tried to obtain government land through the Development Trust of Zimbabwe (DTZ), of which he was Director.[35] One example of this was the widely publicised VIP hunting scandal in 1994. The scandal highlighted the economic importance of wildlife resources and their utilisation, stemming from the fact that the amount of revenue that sport hunting can generate means that, for certain individuals, it is worthwhile attempting to wrest control over those resources away from the Parks Department, Campfire areas and safari operators.

The safari areas in the Parks and Wildlife Estate are set aside to allow low-cost hunting for Zimbabweans.[36] However, until 1995, there was a loophole in the wildlife and hunting laws – a hangover from the settler rule period – which allowed free hunting for VIP visitors in the Parks and Wildlife Estate. In 1994, it was reported that Joshua Nkomo had abused the VIP hunting privilege, in order to allow a group of Austrians access to hunt on Parks and Wildlife Land.[37] Two Austrians, Dr Otto Schreier and Walter Eder, had known Nkomo since the 1970s, and Schreier assisted ex-combatants in the liberation war of the 1970s.[38] In 1994, it was reported that a senior member of the President's Office was assisting Nkomo in identifying commercial hunting opportunities in the former Churundu sugar estates in the Zambezi valley. Nkomo visited Walter Eder in Austria, along with George Pangeti of the Parks Department, to seek investment in the hunting operation in the Zambezi valley.[39] Eder

---

[35] Anonymous interviewee.
[36] Interview with Charl Grobbelaar, Chief Executive, Zimbabwe Hunters Association, 12.2.95, Harare.
[37] *Daily Gazette* 3.6.94 'Abuses of Power'.
[38] *Daily Gazette* 2.6.94 'Nkomo in VIP Hunting Scandal'.
[39] *Financial Gazette* 2.6.94 'Nkomo in Hunting Racket'. Pangeti's presence at the meetings in Austria was never explained.

and Schreier were granted free VIP hunting rights in exchange for investment in Nkomo's hunting ventures, for donating four vehicles to the Parks Department and for building permanent camps and a dam in the area.[40] Yet Mundangepfupfu claimed that the hunts were paid for and approved by the Ministry of Environment and Tourism.[41] The company owned by Schreier, Privilege Safaris, applied for VIP hunting rights in the Makuti/Charara area and in Nyakasanga for a three year period. Whereas previously only three or four permits were issued annually by the Parks Department, that year thirty permits were issued. Those who opposed Nkomo's use of VIP hunting alleged that Privilege Safaris did not pay for the licences to hunt. What did become clear was that the VIP hunters did not adhere to the bag offered by the Parks Department in the Nyakasanga area of the Zambezi valley. It was reported that they overhunted certain species, so that their bag represented the quota for five years,[42] and that unethical hunting practices were employed, such as hanging baits near water, shooting from the roadside, night shooting and hunting female lions.[43] In 1995, the extent of the hunting came to light when reduced hunting bags were sold for the areas used by the VIPs. Nyakasanga saw a 45% drop in earnings from sport hunting, and only five elephant bulls were offered, as opposed to the usual ten. Charl Grobbelaar, of the Zimbabwe Hunters Association, attributed it to the VIP hunting scandal.[44]

While the media and some conservationists pressed for an inquiry, it became clear that investigations into the scandal were blocked. In fact, Graham Nott, the head of the Investigations Branch, threatened to resign if he was barred from looking into the matter.[45] Nkomo threatened to sue Modus Publications for their allegations of his involvement in the VIP Hunting Scandal in the *Daily Gazette* and *Financial Gazette,* claiming that the newspapers were merely running a personal vendetta against him.[46] One source, who opposed Nkomo, suggested the case was dropped because the President's Office allowed the Central Intelligence Organisation to investigate and it was dealt with from there.[47] The scandal highlights the fact that powerful figures have attempted to obtain control over Parks and Wildlife resources, and have been involved in manipulating licences for those resources. Later, Walter Eder was

[40] *Daily Gazette* 20.12.94 'VIP Hunting Axed'.
[41] *Financial Gazette* 9.6.94 'Statements About Nkomo Misleading Erroneous' (Letter from Comrade (Cde.) Tichafa Mundangepfupfu).
[42] Interview with Stephen T. Bracken, Second Secretary, US Embassy, 21.3.95, Harare.
[43] *Financial Gazette* 23.6.94 'No Smoke Without Fire' (Letter from 'conserve the valley'); Daily Gazette 23.7.94 'New Evidence Links Nkomo to VIP Hunting'.
[44] *Financial Gazette* 16.3.95 'Impact of VIP Hunting Begins to Show: Wildlife Auction Flops'.
[45] *Financial Gazette* 15.9.95 'Parks Investigator Offers to Resign'. Nott later claimed that his offer of resignation was nothing to do with the hunting scandal and was based on his age, 51.
[46] *Daily Gazette* 13.7.94 'Nkomo Seeks Urgent Hearing Against Modus'. He was understood to be asking for Z$1 million in damages, but the case was dropped.
[47] Anonymous interviewee.

arrested in Austria for illegally importing ivory and for illegally possessing ivory at his home in Germany.[48] The involvement of VIPs and senior officials in unethical hunting practices damaged the reputation of the Parks Department. In 1995, the Parks Department fought a proposal by the USA to amend its Endangered Species Act, which would have disallowed sport-hunted ivory from entering the USA. The Parks Department and conservationists viewed these markets as essential to their continuing conservation programme of sustainable use.[49] For this reason, scandals related to sport hunting are potentially disastrous. Safari companies that did not deal with VIP hunting were concerned that such scandals would tarnish Zimbabwe's image.[50] Conservationists were concerned that hunting scandals only provided ammunition for those calling for a ban on sport hunting and a wider ivory ban.[51]

The assumption that a conservation agency can act as a technical and apolitical organisation is erroneous. Although conservation is presented as a non-political arena, the levels of patronage and power relationships within the agency with primary responsibility for wildlife conservation in Zimbabwe indicate the intensely political nature of wildlife policy-making. The Parks Department has control over a considerable area of land and substantial resources, and so its policy decisions and strategies for implementation are inherently political and it cannot be immune from wider political battles in the ruling party. The disputes over commercialisation revealed two distinct factions in the Department, which were related to wider divisions in Zimbabwean politics and society. Those who supported it had a long-standing commitment to conservation, whereas opposers feared a loss of power and control over Parks resources. This is closely related to the power struggle in the Parks Department. The two factions can be characterised as those with a professional commitment to conservation and those whose influence is derived from their domestic political reputation. The two alliances effectively used rumour and the media to discredit each other. However, with the removal of a number of key conservationists from the Department, the power struggle appeared to have ended with the patronage alliance left in charge. The power struggle demonstrated the importance of patronage and clientelism in Zimbabwean politics. While the power struggle was cloaked in debates over indigenisation and black empowerment, it was really about factions and their attempts to gain control over resources and power. The final result of the power struggle has been the paralysis of the Parks Department due to financial strains and mismanagement, corruption and a loss of skilled professionals.

[48] Interview with Brian Latham, journalist, 13.7.95, Harare; *Financial Gazette* 16.6.94 'VIP Hunter Bagged'; *Daily Gazette* 15.6.95 'VIP Hunter Under German Probe'.
[49] Interview with Jon Hutton, Director of Projects, Africa Resources Trust, 13.3.95, Harare.
[50] *Daily Gazette* 23.7.94 'New Evidence Links Nkomo to VIP Hunting'.
[51] *Financial Gazette* 22.9.94 'Abuse of VIP Hunting Puts Wildlife Under Scrutiny'.

3

The Use of State Force

in Anti-Poaching & in Poaching

Elephant and rhino poaching in sub-Saharan Africa is one of the key conservation issues that external organisations have focused on. Ivory poaching, in particular, has been a wildlife issue that has captured the imagination of the Western public. However, the image of who is carrying out the poaching and what its root causes are has been structured and organised by international conservation organisations working in conjunction with governments. In fact, poaching is a highly complex problem for conservationists and one that cannot be simply stamped out by increasing the law enforcement capacities of African state agencies. Poaching can be defined as any extractive use of wildlife that is considered illegal by the state, and state-sponsored anti-poaching policies tend to revolve around definitions of what it considers to be legitimate and legal utilisation of wildlife. However, enforcement of anti-poaching policies indicates the possibilities and limitations of state coercion as a means of ensuring wildlife conservation. Anti-poaching has been a consistent theme in conservation policy in Zimbabwe, partly because the need to protect wildlife from certain forms of utilisation has resulted in preservationist legislation, which allows the state to use its coercive powers against those who seek to use wildlife without state permission. The anti-poaching effort formed an alliance between the central state, external donors and international green non-governmental organisations (NGOs). In the campaigns by international NGOs for funds for anti-poaching there is little analysis of the motives for poaching and all poachers are deemed to be equally culpable. Indeed, it is merely the fact that elephants and rhinos were being poached that forms the basis of the fund-raising efforts and justification for anti-poaching methods. The view of commercial poachers that has been presented by NGOs and governments is that they are black, poverty-stricken and hail from neighbouring countries. The public perception in Zimbabwe is that poachers are Zambians and, occasionally, Mozambicans. This view is reinforced by the media and the Parks Department.

The purpose of this chapter is to demonstrate that state coercion is not an effective strategy to tackle poaching, because commercial poaching is internationally organised by powerful sets of interest groups, including the armies and rebel groups of southern African states. In fact, some key elements in governments in southern Africa have been complicit in allowing commercial poaching organisations to continue their lucrative business. Complicity in this case is either through ignoring the military role in poaching for political reasons or by supporting it, again to achieve a particular political objective in terms of regional geopolitics. This chapter indicates that attempts to halt poaching that are not aimed at those organisations that poach and internationally trade in wildlife products are doomed to failure. First, this chapter will analyse the nature of commercial poaching. Secondly, it will examine the extension of the coercive powers of the state through a case-study of the so-called war on poachers, anti-poaching techniques and Intensive Protection Zones (IPZs). Finally, it will investigate the limitations of state powers by examining the levels of official and military involvement in poaching in the southern African region, and this will include an analysis of the role of wildlife in South Africa's destabilisation campaign of the 1970s and 1980s.

Commercial rhino poaching and the extension of state coercion

Peluso argues that a state's capacity to control and extract resources is a function of relations between state and society, but that the state's ability to enforce policy varies widely. Extraction of resources has included allocation of land for tourism and rights of access to wildlife and this has led to the state and local people holding competing claims over national parks and wildlife. As a result, the state has used conservation as a means of coercing local people (Peluso, 1993: 199–202; Neumann, 1997). Successive governments have excluded local people from national parks and outlawed the use of wildlife in order to preserve a national asset for tourism, sport hunting and game viewing, which are mostly the realm of the wealthy. All poachers cannot be considered in the same way, because there are different motivating reasons for poaching. In a number of NGO campaigns in the 1980s, all poachers were treated as equally culpable. The reasons for poaching were not generally explored, since the fact that animals were being poached formed the basis of the campaigns (Peluso, 1993: 205–9). However, subsistence poachers did not present the same level of threat to wildlife that commercial poachers did, and they were not motivated by the same factors.

Subsistence poaching revolves around hunting for food and, in general, it has had a minimal effect on animal populations, because it continues at a low level of intensity and is quite different from commercial poaching. This chapter is primarily concerned with the nature of commercial poaching and the strategies used by state-run wildlife agencies to tackle such illegal hunting. Commercial poaching can be defined as illegal extractive use of wildlife by hunters interested in financial gain, which differs from subsistence poaching, where the primary objective is to obtain meat for consumption. Commercial poachers have tended to target animals that produce valuable commodities, such as ivory, skins, horn, furs and bones. The difficulty with commercial poachers is that they cannot be tackled with the same strategies as subsistence poachers. Communal Areas Management Programme for Indigenous Resources (Campfire)-style policies which rely on commercial incentives to prevent subsistence poaching, would not be sufficient to control commercial hunting. Commercial poachers are unlikely to be convinced of the need for sustainable use and long-term conservation by the community, since they have short-term extractive goals and are not bound by the rules of the local community. Consequently, coercive policies backed by the use of force have been favoured as a means of controlling poaching and, in general, the state has exercised these coercive powers.

Commercial poaching in Zimbabwe has primarily focused on rhinos. The legal trade in rhino horn was banned in 1977 by the Convention on International Trade in Endangered Species (CITES), but, despite this long-standing ban, rhino numbers have continued to decline. While most of eastern and central Africa was concerned with ivory poachers in the 1980s, Zimbabwe experienced a massive poaching onslaught focused on its black rhino. In 1977, there were approximately 65,000 black rhino in Africa; in 1990, there were just 4,000 left (Cumming et al., 1990). Zimbabwe had one of the largest and most secure populations until serious commercial poaching of black rhino began in 1984. It was revealed in 1992, during a Parks Department dehorning exercise aimed at saving rhinos from poachers, that the numbers had declined significantly and that, out of an estimated population of 3,000, only 240–350 remained in the wild, with 150 already moved to privately-owned ranches (DNPWLM, 1992b: 3).

Commercial poaching is partly an extension of the same resistance by local subsistence poachers to exclusionary wildlife policies, which are followed throughout sub-Saharan Africa. However, by definition, commercial poachers hunt for largely economic reasons, so anti-poaching policies that rely on creating local political constituencies in favour of conservation are less likely to be effective. Commercial poachers had no interest in sustainable utilisation of wildlife, but were concerned with immediate financial gain. The earning potential from poaching could be extremely attractive for those who could not hope to earn the same amount in six months or more. Accurate prices for poached rhino horn are difficult to obtain, since the trade is illegal, but one estimate suggests that in 1993 1kg of rhino horn could fetch a poacher US$100 (Milliken et al., 1993). As a result of the international ban on rhino horn trade, range states were not able to derive income from such sales. Instead, poachers, middlemen and dealers captured the economic value of the species being hunted.

A number of factors contributed to the increase in commercial poaching in the 1980s and 1990s. These factors demonstrate that poaching is not a localised issue, but is part of the wider political and economic context of the international system. First, the increase in supply of weapons on the continent had an impact on rates and patterns of poaching, a by-product of civil wars, liberation wars and superpower competition in the 1970s and 1980s in southern Africa (Currey and Moore, 1994: 2; Duffy, 1999). The expansion in the supply of automatic weapons increased the efficiency of poaching rings. Poachers were now able to kill more rhinos and whole herds of elephants in one hunting expedition. In fact, the Minister of Defence acknowledged in 1990 that poaching was 'no longer ordinary poaching, but some kind of military operation'.[1] Poachers were clearly becoming more organised in the late 1980s and they were often better equipped than the Parks Department's own anti-poaching patrols.

Economic and political instability in Zambia were contributory factors in the rise in poaching. Glenn Tatham, in charge of anti-poaching in Zimbabwe, stated that it was the economic situation in Zambia that was at the root of the increase in poaching in the Zambezi valley, and the townships of Lusaka had become recruiting grounds for sending young men to poach rhinos. Zambia experienced massive economic decline in the 1980s and the economic structural adjustment programme compounded the hardship felt by many Zambians. Consequently, young Zambian men were more and more willing to risk their lives for lucrative rhino horn. The image of the poacher that was presented to the public by the Zimbabwean Government and international conservation organisations was of a poor, disorganised Zambian, and poachers were often reported to have escaped to Zambia and dead poachers identified as Zambians.[2]

However, the changing pattern of poacher incursions demonstrated that the received view of poachers as outsiders was not entirely accurate. Poachers were operating further and further into Zimbabwe and, as early as 1988, they were reported to be hunting in Chete safari area and Binga District.[3] This was a serious new development in the rhino wars. Poaching was no longer confined to border areas, with easy access for

[1] Comrade (Cde) Ruchard Hove, Minister of Defence, quoted in *Herald* 3.8.90 'Rhino Poachers Resort to Sophisticated Military Tactics'.

[2] *Herald* 18.5.92 'Greater Effort Needed to Protect Rhino-Warden'; *Herald* 4.1.86 'Four Poachers Killed By Rangers'; *Herald* 9.10.90 'Gang of Rhino Poachers Killed'; see also Milliken et al. (1993: 22–25).

[3] *Herald* 16.5.88 'Three Rhinos Killed as Poachers Move into Hwange Area'.

poachers. This also deprived the Parks Department of the convenient excuse that catching poachers was so difficult because they could easily escape over an international border, where the Department had no juris-diction. Such levels of incursion by poachers raised questions about how poachers could have remained undetected during their long journeys to find the last few rhinos. It became evident that, in order to hunt in Hwange National Park, Chizarira and Chete, the poachers would have to be assisted by Zimbabwean nationals. The Parks Department's own *Zimbabwe Black Rhino Conservation Strategy* noted that poaching was general throughout the country and that the inception of hunting by Zimbabwean citizens was aggravating the situation (DNPWLM, 1992c: 1). It was clear that Zambian poachers could not have continued their onslaught with such ferocity without the complicity of Zimbabwean nationals – the assumption that all poachers were outsiders was called to question.

Local rural people were not the only ones who assisted Zambian poachers in finding rhinos or selling their horns. The lucrative trade in wildlife products fused the triangle of interest groups – politicians, business people and diplomats – that were regarded in conservation circles as responsible for the poaching increase in Africa.[4] The importance of the diplomatic community as contributors to the rise in poaching in the 1970s and 1980s was also highlighted. The increase in automatic weapons came at a time when new diplomatic missions were being opened all over Africa after decolonisation, a factor that was particularly important in the case of the increase in Asian embassies on the continent, which acted as conduits for ivory and rhino horn, which were then sent to Asian consuming states. One interviewee, who was involved in the conservation community in Zimbabwe, stated that South Africa was still the biggest port of exit for illegal ivory, which was moved out in crates or in diplomatic bags, which could not be opened. The expansion of the diplomatic community also coincided with the opening up of transport connections in sub-Saharan Africa. Trucking companies and airlines were used by poachers and smugglers to shift ivory and rhino horn out of the continent. In evidence to the Kumleben Commission in South Africa, one South African Defence Force (SADF) Special Forces Officer stated that while he was gathering information on anti-government forces in Zambia, Namibia and Botswana, he was informed that these groups were using wildlife products to finance their operations, and he stated that several cargo companies and drivers were involved in smuggling ivory, rhino horn and precious stones.[5] One interviewee noted that, although there was no single Mr Big behind poaching in southern Africa, there was a belief that there were probably three or four people in Lusaka, Johannesburg and Harare who were politically protected and who were ultimately responsible.

[4] Anonymous interviewee.
[5] *Mail and Guardian* (South Africa) 8.9.95 'New Claims of SADF Ivory Smuggling'.

Anti-poaching policies

The increase in poaching demanded formulation of stricter anti-poaching strategies to protect targeted wildlife populations. Peluso argues that governments in sub-Saharan Africa have joined in an alliance with international NGOs over anti-poaching, which has allowed the state to expand its powers over marginal social groups that live in areas near prime wildlife habitats (Peluso, 1993: 205–9; Neumann, 1997). Anti-poaching policies extended the state's coercive reach into regions that it had not previously targeted. The relatively remote and sparsely populated parts of the Zambezi valley became a particular focus for policies designed to protect wildlife from illegal use. These coercive strategies revolved around the notion that it was the state's right to protect a national asset (or even an international asset) against illegal and unsustainable utilisation. The idea that coercion was the state's right in turn led to the violation of individual human rights, when poachers were shot on sight. Anti-poaching is a preservationist policy response to illegal hunting, being an attempt to protect targeted animal populations from further use that is not sanctioned by the state. The perceived need to protect wildlife from human interference and human use resulted in policies that required strict separation of the human and animal worlds.

The so-called war on poachers in Zimbabwe was carried out at the same time as the internationally infamous poaching wars in Kenya. It was a highly controversial policy, which attracted condemnation from a number of international human rights organisations, because it relied on increasing the state's powers of enforcement to stamp out illegal hunting. Operation Stronghold was formulated in 1984 as the centre-piece of the war on poachers, and was intended to strengthen the anti-poaching capacity of the Parks Department. It relied on four dimensions: local reaction, national reaction, international reaction and intelligence gathering, which included communication with the international media, provision of support and adequate funds from central government, anti-poaching and rhino translocation, and finally intelligence on the international rhino horn trade (Tatham, 1986: 1; DNPWLM, 1992c: 28). The need to expand the powers of patrols was even more urgent when it became clear that poachers were ultimately controlled by organised gangs operating from North Africa to the Far East and that rhino poaching in the 1980s presented the Parks Department with a completely new situation.[6] In response, in 1987 crack units were set up in each of Zimbabwe's eight provinces and a further ninth unit was to be available in times of emergency to assist provincial units.[7] In addition, aircraft were used to strengthen the operations, because it was hoped that spotter planes would act as a deterrent.

---

[6] Interview with Glenn Tatham, Head of Operations, Department of National Parks and Wildlife Management (DNPWLM), 10.5.96, Harare.
[7] *Herald* 11.2.87 'Crack Units Set Up'.

The core of Operation Stronghold was the so-called shoot-to-kill policy. During the years of rhino warfare in Zimbabwe, the Parks Department has arguably sustained one of the most systematic and deadly anti-poaching efforts any country has undertaken (Milliken et al., 1993: 1–3). The head of anti-poaching operations, Glenn Tatham, justified the use of shoot-to-kill (though he objected to the use of the term as sensationalist) as the last resort decision of the ranger on the ground. In clashes with poachers, rangers were faced with the situation of having to defend themselves and, in so doing, poachers sometimes lost their lives. He claimed that rangers and their families had been followed and threatened when they returned to their homes or Parks headquarters in Harare. Tatham was concerned to explain that the rangers were carrying out their duties and protecting the national heritage in the same way that police would protect a bank vault against armed robbers.[8]

In contrast, *The Times of Zambia* accused Zimbabwean game wardens of being trigger-happy,[9] while Bonner argues that white Zimbabwean rangers seemed to relish the war (Bonner, 1993: 13–24). This was partly because the war on poachers began only four years after the end of the liberation war, and a number of white Parks officers who had served in the Rhodesian Army found themselves together once again to combat a new enemy.[10] The shoot-to-kill policy obtained Mugabe's assent in 1986, as a consequence of which, between 1984 and 1993, approximately 170 poachers lost their lives, nearly all Zambians, except for a few Mozambicans and Zimbabweans (Milliken et al., 1993: 1–3). In 1987, the then Minister of Natural Resources, Victoria Chitepo, told Parliament that the anti-poaching campaign was now classified as a full-scale war.[11] Similarly, while addressing the House of Assembly in 1987, President Robert Mugabe redefined clashes between poachers and rangers as 'contacts', so that poachers were then treated in the same way as Mozambique National Resistance (Renamo) bandits from Mozambique, thereby legitimising the state's deployment of the security forces 'in full force'.[12] Nineteen ninety was the first year in which more poachers were killed than rhinos.[13] This news was welcomed in conservation circles and it was thought that the Parks Department was winning the rhino wars at last.

In operating a coercive strategy, the Parks Department found that it was open to legal challenges concerning the execution of suspected poachers. As a result, the Parks Department was keen to provide legal protection for

[8] Interview with Glenn Tatham, Head of Operations, DNPWLM, 10.5.96, Harare.

[9] *Herald* 10.1.86 'Poachers Beware'.

[10] Interview with Glenn Tatham, Head of Operations, DNPWLM, 10.5.96, Harare.

[11] *Herald* 9.4.87 'Full Scale War on Poachers Declared'.

[12] *Herald* 5.11.87 'Parks Units in Raging Battle'; see also *Herald* 9.2.87 'Two Killed as Anti Poaching War Hots Up'.

[13] *Herald* 18.7.90 'Poachers Beware as Parks Win the Rhino War'; *Herald* 21.1.89 'Top Ranger Murdered by Poachers'; *Herald* 22.1.89 'Open Warfare'.

its staff. Part of the shoot-to-kill strategy was to ensure that Parks rangers were able to carry out their duties without fear that they would be charged with murder. After some debate, the Protection of Wild Life (Indemnity) Act was passed in 1989. The Act meant that no criminal liability was attached to any indemnified person for any action carried out in good faith, in connection with the suppression of poaching. Fortunately for the accused officers, it applied to anti-poaching actions prior to the introduction of the Act. Some MPs were concerned that the Act could be used by Parks staff to protect themselves when innocent people were killed.[14] Nevertheless, the new Act provided a boost for morale in the anti-poaching units. Indemnifying Parks officials was a low-cost option for the government, since it gave the impression that anti-poaching and wildlife were priorities, while not requiring significant levels of expenditure or commitment from government.

However, human rights organisations viewed the Act as inhumane and in violation of basic human rights. Although this view is generally unpopular in Europe and the USA, where protecting wildlife from poachers is presented as a global good, it is forgotten that so-called poachers are routinely killed and, never having been prosecuted under the law, they are denied a fundamental right to be considered innocent until proved guilty. It was clear that shoot-to-kill placed more value on the rights of endangered species than on the rights of alleged illegal hunters. Still, animal rights and deep-green organisations, such as the US-based Friends of Animals, applauded shoot-to-kill. This highlighted the competing claims to rights, including the rights of the state to protect a national (or even international) asset, human rights and the rights to life of the poacher and, finally, the rights of the animals being poached. Amnesty International called for the Indemnity Act to be repealed and suggested that some of the poacher deaths during the rhino wars should be redefined as extrajudicial executions, since victims were only suspected of poaching and their involvement was never proved under the law (Amnesty International, 1992). Human rights and development organisations deplored the loss of life in the war on poachers. The concerns that were raised by ZANU-PF MPs that the law could be used to protect those involved in murder or extrajudicial execution were echoed by Amnesty International. The numerous deaths of poachers were condemned by Amnesty International, and a report by the organisation questioned the legality of the deaths and raised concerns that rangers were able to shoot suspected poachers with impunity (Amnesty International, 1992).

The war on poachers deeply affected regional relations; it severely strained the relationship between Zimbabwe and Zambia, as Zambian MPs publicly condemned the killing of poachers and urged their own government to take measures. However, President Kaunda was aware of

---

[14] *Herald* 1.6.89 'Stormy Passage for Wildlife Bill as Backbenchers Revolt'.

the need to retain good relations with Zimbabwe, and he did not condemn the killings but eventually backed the war on poachers. His government pledged to cooperate with Zimbabwean authorities in arresting known poachers and promised patrols on the Zambian side of the Zambezi valley.[15] One of the most important themes in anti-poaching was the continual attempts to obtain cooperation from the Zambian government and wildlife authorities. The Parks Department had been promised increased cooperation from the mid 1980s, but it was not until the 1990s that concrete assistance came, when anti-poaching patrols were initiated on the Zambian side of the Zambezi river.

Concrete support and assistance for the Parks Department were essential elements in the success of anti-poaching measures. As Parks staff were facing heavily armed poaching gangs, it was critical that they had adequate equipment and funding. One of the goals cited in the *Short and Medium Term Action Plans for Black Rhino* (DNPWLM, 1992b) was to adequately equip the anti-poaching patrols. However, the central government did not provide the Parks Department with the levels of support necessary to carry out the ambitious programme set out in Operation Stronghold. Complaints about lack of funds surfaced as soon as Operation Stronghold began. It was estimated that effective *in situ* protection for the black rhino would cost US$400 per square kilometre (Milliken et al., 1993: 27). Yet, in general, wildlife enforcement budgets have fallen in Zimbabwe and, from 1988, annual budgets suffered dramatic declines, with a drop in operational expenditure (per square kilometre) from US$24 in 1988 to US$ 2.63 in 1993 (Dublin et al., 1995: 14). This was far short of the necessary levels of expenditure needed to ensure that law enforcement was effective. The cuts in funding from central government have had a crippling effect on the Department.

In the absence of adequate government funding, the solution suggested by the Parks Department was to have AK-47s on loan from the Zimbabwe National Army (ZNA), while the Department aimed to build up its own stocks (DNPWLM, 1992b: 16–21). Scouts were also expected to use second-hand guns from the armed forces, which often failed in contacts with poachers. Underfunding had an adverse effect on staffing levels as well, with only 173 rangers to cover the lower Zambezi valley, an area of 31,000 square kilometres.[16] To boost morale, the Department set out an incentive scheme. It gave examples of rewards that staff could expect for killing or capturing poachers, including Z$10,000 for the capture and arrest of a whole gang of three or more, Z$4000 for killing a whole gang, Z$500 for killing one poacher, Z$10 for finding a carcass and Z$5 for each spent cartridge found (DNPWLM, 1992b: 14). This raised ethical questions about placing prices on the deaths of poachers in contacts.

---

[15] *Herald* 9.1.86 'Kaunda Backs Zimbabwe on Poaching'; *Herald* 21.2.86 'Zambian MPs Hit Out at Poacher Killings'.
[16] *Sunday Gazette* 18.7.93 'Parks Dept Needs More Funds to Win Poacher War'.

Nevertheless, the anti-poaching effort was severely hampered from the start by the failure to provide adequate equipment and funds. Morale was reduced as game scouts and rangers began to question whether it was worth risking their lives in contacts with poachers without the necessary equipment and for such poor wages. The failure to allocate adequate funds for anti-poaching demonstrated that wildlife was not a high priority. With competing demands for funds for development projects, education and the armed forces, wildlife lost out. The failure to allocate adequate funding to the Department for the rhino wars cast doubt over the government's commitment to beating the poachers in the Zambezi valley.

The lack of funding from central government caused the Parks Department to look to the international conservation community and to donors for assistance. State conservation agencies in sub-Saharan Africa have actively courted alliances with international donors and NGOs in pursuit of funds and support for their anti-poaching policies. The relationship between international NGOs and the Parks Department can be viewed as part of the increasing privatisation of relations between Africa and the international system (Princen and Finger, 1994; Clapham, 1996). NGOs formed a link between local conservation efforts and global organisations, with their access to funds. Donors and NGOs have been visibly supportive of Zimbabwe's anti-poaching campaigns. For example, anti-poaching units were given a Bedford truck with a fax machine by the British Government in 1992, and the US Department of the Interior gave US$104,000 specifically for anti-poaching in the Zambezi valley.[17]

NGOs sought to raise their own public profile by being associated with the fight to save the elephant and rhino from poachers. This was also a lucrative association for NGOs, since many funds were raised specifically for anti-poaching. The image of the poacher used by these campaigns was that of the poverty stricken, yet greedy, outsider robbing the host countries of their natural heritage. It drew on the Western belief that wildlife had to be protected from indigenous peoples to prevent it being hunted to extinction (Neumann, 1997). Local and international charities and NGOs were the most prominent donors, regularly providing much needed funds and equipment. This was presented as the only concrete way of assisting governments and wildlife authorities. For donors that wished to be seen to be doing something about poaching, provision of equipment and funds served that purpose. The Department's reports and newspaper reports document numerous donations of equipment including planes, helicopters, boats and vehicles. For example, the International Black Rhino Foundation donated a helicopter fitted with a Global Positioning System (GPS) in 1993.[18] What was signif-

---

[17] *Herald* 26.3.92 'New Equipment to Help Fight Poaching Menace'; Ziana report 19.3.91.
[18] *Herald* 6.2.93 'National Parks Receives Helicopter Worth $3m'.

icant about the types of donation from international NGOs was that they were all large donations of capital equipment, which were high-profile and easily recognised. The NGOs could raise their public profile by being associated with the anti-poaching effort. However, the patrols were also short of basic equipment, such as sleeping bags, firearms and even rations, but very few organisations donated these, because they did not attract the same level of publicity as donations of trucks or aircraft. The constant complaints about lack of funds for anti-poaching were more to do with how the Department and the Ministry of Environment and Tourism decided to spend their allocations from the central treasury. Indeed, the government was aware that NGOs were willing to provide assistance and, consequently, the government felt it was not necessary to provide adequate funding for anti-poaching themselves.

In addition, NGOs that otherwise would not be seen to be supporting violence have turned a blind eye to the activities of a number of conservation agencies. In turn, NGOs have justified their position on the basis of preserving biodiversity (Peluso, 1993: 199–200). This was assisted by a simplistic definition of rangers as wildlife protectors and poachers as wildlife destroyers. One of the most controversial and politically-loaded donations by an NGO came from the World Wildlife Fund (WWF) International, which donated a helicopter for the anti-poaching effort in 1987. The helicopter was used in the shoot-to-kill policy, and it was trumpeted in a Zimbabwean newspaper as the helicopter used to shoot a poacher in the Sapi safari area.[19] However, the helicopter quickly turned into a public relations disaster, as WWF-International became embroiled in a human rights versus animal rights row. The provision of the helicopter was severely criticised by development and human rights organisations for its role and obvious support for shoot-to-kill. WWF-International denied it was ever intended to be used as a helicopter gunship, but Bonner argues that there was fierce internal debate about its use as a gunship (Bonner, 1993: 77). Nevertheless, the policy of WWF is not to provide weaponry or funding for weaponry. The Z$55,000 needed annually to run the helicopter was eventually withdrawn. Publicly, WWF-International stated that their decision was due to lack of funds and because they did not agree with Zimbabwe's conservation policy, specifically culling.[20] However, it was suggested that the helicopter was withdrawn because of international criticism of WWF's involvement in shoot-to-kill. From this case, it was clear that, although NGOs could use their association with anti-poaching to raise their own profile in a positive manner, it could also be disastrous.

In the face of continued large-scale rhino poaching, it was obvious that Operation Stronghold did not have the desired effect. Poaching consti-

---

[19] *Herald* 14.3.88 'Swiss Bid to Check Rhino Slaughter'; *Herald* 16.5.88 'Eighth Poacher Shot Dead'. The helicopter was originally given by the Goldfields Trust to the Duke of Edinburgh in his capacity of WWF-International's President, and then it was assigned to Zimbabwe.
[20] *Financial Gazette* 31.3.89 'WWF to Halt Copter Aid'.

tuted a threat to the authority of the state as a wildlife manager, because it was a clear and very public form of disobedience to the state's legal framework, and demonstrated the limits of its law enforcement capabilities. Under these circumstances, the state resorted to a greater level of violence and coercion to exercise control over potentially lucrative wildlife resources. In 1993, Operation Safeguard Our Heritage (also known as Operation Save Our Heritage) was launched to provide assistance for flagging anti-poaching activities. It involved deploying special police and army support units to assist the Parks staff. It was a controversial policy decision, because pro-development organisations, such as Africa Resources Trust, complained that further protectionist measures would fail as previous ones had done, and the only answer was to legalise the rhino horn trade and harness the revenue for rhino conservation.[21] It was also controversial because those who had exposed the military involvement in poaching in the south east of Zimbabwe were concerned that this decision merely afforded elements in the military more opportunities to poach.[22] Operation Safeguard Our Heritage also damaged morale, when it became clear that the army and police support units were being paid far more than Parks staff for carrying out the same duties: in 1993 police and soldiers received Z$90 per day bush allowance while game scouts received just Z$11.[23]

Operation Safeguard Our Heritage provided the Zimbabwean government with a new role for the armed forces. For politicians, the opportunity to redeploy the Defence Forces was timely, since the establishment of the Special Support Units coincided with Zimbabwe's withdrawal from guarding the Limpopo and Tete Corridors in Mozambique. With the peace agreement in Mozambique, it was no longer necessary to devote such large numbers of troops to keeping vital supply routes open, particularly the route to Beira. This presented the problem of what to do with the demobilised soldiers, and conservation provided the solution. The skills obtained in military training are similar to those required for field conservation, including an ability to survive in wild and remote places, knowledge of flora and fauna, ability to use and maintain weaponry and military-style planning and ambushes. The military in both Zimbabwe and South Africa have offered their services to conservation as part of their search for a new role in the post-apartheid era (Ellis, 1994: 64). Operation Safeguard Our Heritage resolved a number of problems, as it could be heralded as proof of the state's renewed commitment to rhino protection, it could considerably strengthen the anti-poaching effort and it also solved the problems associated with demobilisation of the military.

---

[21] Interview with Jon Hutton, Director of Africa Resources Trust, 13.3.95, Harare; see also *Financial Gazette* 12.5.95 'Call to Arms Not a Solution' (letter from Jon Hutton).

[22] Anonymous interviewee.

[23] *Sunday Gazette* 18.7.93 'Parks Dept Needs More Funds to Win Poacher War'.

Preservationist forms of protection were the underlying principles of Operation Safeguard Our Heritage; yet, for Zimbabwe's wildlife managers, preservation merely formed the centre-piece of a crisis management strategy, rather than being an expression of a strict policy line. IPZs were established as part of Operation Safeguard Our Heritage. IPZs can be defined as special areas within the Parks and Wildlife Estate where rhino populations are monitored and live in a situation of vastly increased law enforcement (Pearce, 1991: 91–92; DNPWLM, 1992b: 4). The IPZs are concentrated in the national parks around the Zambezi valley, with Matusadonha National Park as the main IPZ. In effect, IPZs created micro-parks within national parks. For example, Sinamatella Camp in Hwange National Park was designated as a black rhino IPZ. In the long term, the Parks Department has promoted sustainable utilisation of rhinos, through sport hunting and sales of rhino horn, as the key to future conservation. However, once faced with an emergency situation, preservationist approaches were emphasised, such as removal of rhinos for breeding programmes in the USA and Australia and the transfer to IPZs. It was hoped that these measures would protect rhinos from poachers in the short term, in order to permit future utilisation when rhino numbers increased (DNPWLM, 1992b,c). A second crucial element in the IPZ strategy was dehorning because it was hoped that the removal of the valuable part of the rhino would deter the poachers (DNPWLM, 1992b: 42). However, rhinos continued to be hunted, because poachers could not determine whether a rhino had a horn or not in the thick bush, and it was also discovered that dealers in the Far East were gambling on the extinction of the rhino to increase the value of their stockpiles.[24] Dehorning was an enormously expensive exercise, the Parks Department estimated that it cost between US$350 and US$1,800 per animal, depending on the terrain, population density and accessibility (Milliken et al., 1993: 48). Consequently, the initial round of dehorning in Hwange in 1991 had to be suspended, due to lack of funds. In fact, it was reported that most of the overseas donor funds for dehorning had been spent on running a helicopter to spot rhinos, hired from Botswana.[25]

One interviewee commented that the rhino poaching situation had indeed stabilised, but for no other reason than the sheer paucity of rhino numbers – after years of poaching, there were very few left.[26] In the *Zimbabwe Black Rhino Conservation Strategy*, the severity of the rhino wars in 1992 and the need to organise staff on military lines to combat poaching were highlighted (DNPWLM, 1992c: 1). One interviewee stated that, in the area he hunts, a WWF aerial survey was carried out that found twenty-five carcasses of elephants and other animals, but the official

[24] Interview with John Gripper, Sebakwe Black Rhino Trust, 14.10.94, Ascott-Under-Wychwood.
[25] *Herald* 14.11.91 'Over 230 Rhino in Hwange Dehorned in Trial Programme'.
[26] Anonymous interviewee.
[27] Anonymous interviewee.

figure was five.[27] Yet poaching continued to be reported as a serious threat at the same time as reports that the rhino wars had been won appeared. Nevertheless, Dick Pitman of the Zambezi Society suggested that IPZs had ultimately resulted in a stabilisation of the poaching situation in the Zambezi valley.[28] What is clear is that, for whatever reason, the rates of rhino poaching declined after a high in the late 1980s. Fewer rhinos were killed from 1993, but that may be because there were only a few left. The rhino wars demonstrated the links between conservation and the coercive powers of the state that were used against Zimbabwean nationals and against poachers from neighbouring countries. This was clearly underlined by the redeployment of army and air force units to conservation and internal security, once their role in regional and national security was diminished. The extension of the coercive powers of the state through appeals to conservation did prove to be politically effective against this particular form of commercial poaching, but these coercive strategies failed to address the widescale poaching undertaken by liberation movements in the region and by arms of the state in southern Africa.

Ivory poaching

Although the coercive powers of the state could be readily deployed against poachers in and around the Zambezi valley, the poachers in the south-east lowveld proved to be more difficult to deal with. Poaching that is carried out by arms of the state has proved to be very problematic, and this is because a junior arm of the state, such as the Parks Department, is ill-equipped to deal with more powerful state agencies, such as the army. Ivory poaching in Zimbabwe reveals that commercial poaching is a highly-organised international business, which is reliant on structures within a state or society, rather like any other business. It is clear that militaries and rebel groups in southern Africa were the organisations with the necessary capacity to hunt with relative impunity and with the internal structures and criminal links to poach and ship ivory to the consuming states of the Far East. The case of ivory poaching indicates that illegal users of wildlife drawn from outside the state apparatus can be dealt with by coercive arms of the state, whereas illegitimate wildlife users drawn from within the state's own apparatus constitute a greater challenge to state wildlife agencies. In fact, it demonstrated that the Zimbabwean state lacked the internal legitimacy and capacity to stamp out poaching in the south eastern part of the country. The coercive capacities of the Parks Department also proved to be no match for well-armed and funded military forces in neighbouring countries, which were exploiting opportunities for commercial poaching.

[28] Interview with Dick Pitman, Zambezi Society, 30.5.95, Harare.

Although the Parks Department has been largely concerned with rhino poaching, illegal hunting of elephants in the 1980s is a closely related issue. It is an issue that is often ignored, and has even been subject to cover-up and denial by conservation authorities. Ivory poaching has been a problem since the early 1980s; yet its extent was left uninvestigated until a decade later, since when an official definition of who these ivory poachers were has not been forthcoming. However, elephant poaching was more openly discussed as a result of the ban on international ivory trading in 1989. Before and after the ban, the Parks Department consistently argued that an ivory ban would cause an increase in poaching. The reported rise in ivory poaching was deployed as a politically important legitimating argument for Zimbabwe's policy stance on the ivory trade. A number of interviewees cited the ivory ban as the main reason for the increase in poaching, suggesting that, as a result of the ban, ivory prices had at first collapsed but have steadily grown since 1989.[29] Usually, the price paid to poachers was a fraction of the value of the ivory or rhino horn obtained by smugglers and dealers. However, it is very difficult to obtain accurate prices for ivory and rhino horn in the post-ban period, because all trade is illegal. Nevertheless, in the period 1992–5, the trend for black-market ivory prices was upwards (Dublin et al., 1995), despite the fact that the ivory ban was supposed to have effectively shut down such black markets. Parks Department reports demonstrate that Zimbabwe has definitely witnessed a rise in illegal hunting of elephants for ivory. From 1989 to 1992, there was an increase in ivory recovered from poachers by Parks staff and in elephant-poaching activity (Murphree, 1992). In the first six months of 1993, forty poached elephants were found in the Zambezi valley, the previous site of the rhino wars.[30]

There were a number of explanations for this rise. In a study by the International Union for the Conservation of Nature (IUCN) African Elephant Specialist Group, a number of factors were cited as influential in the increase in poaching for ivory in Zimbabwe. Ivory poaching has been linked closely with rhino poaching. As rhino poaching moved into southern Tanzania, Zambia and northern Zimbabwe, a time-lagged increase in elephant poaching followed (Dublin, 1994b: 28). As rhino numbers declined, poachers increasingly turned to ivory to finance their expeditions, because, if they could not return with rhino horn, they could at least recoup some of the cost with ivory. It was clear that the rise in ivory poaching was not wholly due to an increased interest in ivory *per se*, but was a second choice used to finance poaching expeditions in search of other prey, namely the rhinos. Dick Pitman of the Zambezi Society

---

[29] For example, interview with Jon Hutton, Director of Projects, Africa Resources Trust, 13.3.95, Harare; interview with Steven Kasere, Deputy Director of the Campfire Association, 14.3.95, Harare; interview with Charl Grobbelaar, Chief Executive of the Zimbabwe Hunters Association, 12.2.95, Harare.

[30] *Herald* 18.6.93 '4 People Killed During Battles with Poachers'.

suggested that the increase in ivory poaching in the Zambezi valley was also directly related to the decline in money available for law enforcement.[31] Donations for anti-poaching declined after the ban, reflecting the donor belief that the ivory ban was the solution to the poaching problem. In addition, redirection of donor funds to the rhino had a negative effect on funds for elephant range states that were not rhino range states (Dublin et al., 1995: 84–7). Budget declines were linked to a drop in the number of patrols and vehicles available for operations.

While anti-poaching is a classic case of the state's desire to use coercive measures for conservation, poaching also revealed the network of official complicity and corruption in illegal hunting. The involvement of officials in poaching and smuggling is significant, because it neutralises the state's capacity to enforce its own laws. It is clear that the levels of poaching experienced across the continent could not have occurred without a degree of official complicity. At worst, there was direct and personal involvement in sponsoring poachers and conducting smuggling, while, at best, there was official complacency about poaching and turning a blind eye to illegal activities. As a result of this complicity, traders and dealers bought off officials to ensure that their operations were left unhindered. Although there is very little written evidence of official and military involvement in poaching and smuggling, a number of interviewees stated what they perceived to be the case. These included government officials in the Parks Department and others who were directly involved in conservation. In one sense, it does not matter whether their assertions are provable or not, as what is important is that policies and wildlife management are partially based on the assumptions made about poaching and smuggling operations. In addition, anecdotal and oral evidence of such activities are readily accepted in sub-Saharan Africa, where government complicity is unlikely to be exposed in the media or other forums (Ellis, 1989: 321–30).

What is clear is that poaching is part of a vast industry, with a number of different types of poachers, ranging from local subsistence poachers to ruthless commercial poachers allied to governments and illegal smuggling rings (Iain Douglas-Hamilton, cited in Wildlife Society of Zimbabwe, 1993). Illegal trading routes and groups are crucial factors in the survival of the black-market trade supplied by poachers. Smuggling routes for ivory through South Africa, Mozambique and Zambia were almost an open secret amongst conservationists in Zimbabwe. Smugglers are the link in the chain that transports ivory from the field to markets. Available information showed that an illegal trade in ivory was flourishing in one of the neighbouring countries in the post-ban period.[32] Zambia has a busy and lucrative black market in ivory and Lusaka was a

---

[31] Interview with Dick Pitman, Zambezi Society, 30.5.95, Harare.

[32] Glenn Tatham, quoted in *Herald* 18.6.95 '4 People Killed During Battles with Poachers'; see also *Herald* 6.5.95 'Poachers Slaughter Hundreds of Jumbos'.

major conduit for rhino horn, which was mostly poached in northern Zimbabwe (Cumming et al., 1990: 56). Ivory poached in Zimbabwe made its way to Zambian carving houses and from there to South Africa (EIA, 1992: 14). Poaching and smuggling are organised by the same gangs as those involved in drugs, arms and gems, and these gangs operated an international network that supplied poachers and sold ivory. Indeed, in an investigation in the run-up to the ivory ban, the British-based Environmental Investigation Agency (EIA) noted that they were under no illusion that such a task was easy, as there were too many influential people in Africa that had vested interests in preserving poaching (Thornton and Currey, 1991: 119).

During the apartheid era, South Africa actively used ivory and rhino horn poaching and smuggling as part of its programme of destabilisation of front-line states in the 1970s and 1980s. Destabilisation can be defined as the use of military and economic weapons against the front-line states, aimed to create a fundamental shift in their policies towards apartheid, and this was to be achieved by attacking neighbouring states in areas where they were vulnerable (Hanlon, 1986: 27–30). One area of vulnerability was protection of endangered species from poaching. The central importance of Johannesburg in the illegal wildlife trade assisted the policy of destabilisation. Conservation policies and wildlife have had a significant role to play in regional relations in southern Africa. South Africa was the favoured conduit for the illegal imports from the front-line states because it legally permitted imports of ivory (Ivory Trade Review Group, 1989: 11). It was easy to mix legal and illegal ivory until it all entered through the legal ivory trading system. In addition, the experience of sanctions under apartheid provided smugglers with the necessary skills to conduct an illegal ivory trade, and hence the routes used by sanction-busting businessmen were also used to bring ivory into South Africa (Currey and Moore, 1994: 1–10).

It became clear in the 1990s that SADF used these sales to finance the União Nacional para a Independencia Total de Angola (UNITA) and Renamo operations in Angola and Mozambique, respectively. In the 1980s and early 1990s, such involvement was denied and investigations were reported to have been blocked by high-ranking officials.[33] In fact, rumours of military involvement were sufficient in 1988 for the Roos commission of inquiry to be set up, which concluded that only 500 tusks confiscated from UNITA were handled by the South African military.[34] However, the Kumleben commission of inquiry in 1995 found that the Military Intelligence Department was heavily involved in the sale of elephant tusks from Angola between 1978 and 1986 and it described the Roos commission as a charade.[35] It was SADF involvement in the

[33] *Independent (UK)* 23.3.92 'SA Army's Ivory Trail Exposed'.
[34] *Mail and Guardian* 19.1.96 'Report Takes WWF to "Tusk"'.
[35] *Mail and Guardian* 19.1.96 'Report Takes WWF to "Tusk"'.

poaching and smuggling of ivory and rhino horn in Angola that received the greatest attention. The evidence of a high-ranking South African military officer was critical in motivating the investigations. Colonel Jan Breytenbach stated that certain elements in the Chief of Staff (Intelligence) (CSI) were using a front company, Frama Inter-Trading, to clandestinely export teak, ivory, gems and drugs from Angola, and that the purpose of such exports was to finance UNITA and South African operations (Reeve and Ellis, 1995: 227–43; Duffy, 1999: 111).

Another SADF proxy, Renamo, used wildlife poaching as part of its strategy. Renamo was responsible for large scale poaching in southern Mozambique and Zimbabwe. Ivory was used to finance operations (Ivory Trade Review Group, 1989: 11). After Zimbabwe's independence, South Africa took over the running of Renamo and used elephant and rhino poaching by Renamo and the SADF as part of a campaign to destabilise the region. It was reported that Zimbabweans found illegally in South Africa were made to poach by SADF in return for safe passage home (EIA, 1992: 17–20). Such levels of poaching in south-eastern Zimbabwe kept Zimbabwean troops occupied in Gonarezhou National Park, a frozen zone under military command. It was suggested that the South Africans were behind poachers infiltrating Gonarezhou as part of a campaign to create enmity between the governments of Zimbabwe and Mozambique. The MP for Chiredzi South stated that he believed that the poachers were not Mozambicans at all, but SADF masquerading as Mozambicans.[36] Renamo paid for SADF assistance in whatever raw materials they were able to obtain, which included ivory and rhino horn (Thornton and Currey, 1991: 224; Ellis, 1994: 56–59).

Overseas conservation organisations were also complicit in apartheid South Africa's use of wildlife for destabilisation purposes. An undercover scheme, named Operation Lock, to expose the big names behind poaching in Africa was launched by KAS Enterprises in 1987 in South Africa. KAS Enterprises was run by the late David Stirling, founder of the British Special Air Service (SAS).[37] Indeed, the mercenaries that KAS employed were ex-British SAS. They trained anti-poaching units in Namibia and also Mozambicans inside South Africa (Bonner, 1993: 81). Ian Crooke, leader of the anti-poaching team, stated that the operation was funded by wealthy individuals from overseas, although he could not name them.[38] However, other sources suggest that Prince Bernhard of the Netherlands was one of the major funders of Operation Lock; he was also former President of WWF-International and remained closely linked with the organisation. In addition, John Hanks, then Director of Africa Programmes for WWF-International, was a funder and supporter. Britain's Prince Philip ordered an investigation into the affair and it

[36] *Herald* 21.6.88 'Pretoria Behind Poaching Says MP'. The MP was Titus Maluleke of Chiredzi South.

[37] *Herald* 10.9.91 'Plan to Expose Poaching Activities in Africa Flops'.

[38] *Herald* 11.7.89 'SA Undercover Operation Against Poaching in Doubt'.

found that no WWF funds had gone to Operation Lock. The scheme made little progress and in 1989 KAS Enterprises went out of business.[39]

In the context of the aggressive stance of apartheid South Africa towards its neighbours, conservationists in the front-line states were immediately suspicious of the motives of KAS Enterprises, largely due to the fact that they were based in South Africa. Rowan Martin was approached by Crooke, but he declined to cooperate with him because Crooke was vague about his sponsors and the objectives of the mission. Martin suggested that Crooke seemed more interested in military technology than in wildlife conservation.[40] It later became clear that the organisation had been infiltrated by South African intelligence officers. An internal KAS document presented to the Kumleben Commission gave details of how Ian Crooke offered to help South African intelligence monitor anti-apartheid movements in the front-line states and overseas in return for government assistance in anti-poaching and sting operations against wildlife product dealers.[41] In addition, South African intelligence officers were aware that Operation Lock had the potential to uncover SADF involvement in ivory smuggling, and those that infiltrated KAS Enterprise obtained agreement that the SADF and Operation Lock would avoid contact with each other (EIA, 1994: 17). It is clear that wildlife products have had a critical impact on regional relations, because ivory and rhino horn have provided finance for military forces involved in regional wars, and also poaching and anti-poaching formed central elements in South Africa's destabilisation programme. Both of these activities placed a strain on the front-line states in terms of allocation of funds and personnel to deal with poaching. Poaching neatly fitted in with the basic principles of destabilisation, which targeted the use of military and economic weapons against the front-line states.

The implication of Parks staff and the ZNA in poaching and smuggling is also a highly controversial topic in Zimbabwe. Official complicity in poaching and military involvement reduces the state's capacity as a wildlife manager and protector. Anti-poaching is rendered ineffective if the poachers are assisted by the very body responsible for preventing access to wildlife. This is particularly important, given Zimbabwe's policy stance on the ivory trade, concerning which the Parks Department has argued that it was capable of conducting a legal and controlled trade and of ensuring that elephants were not poached and depleted. The implication of officials and military personnel has cast doubt on the ability of the Department to carry out such a policy effectively. Yet a

[39] *Herald* 10.9.91 'Plan to Expose Poaching Activities in Africa Flops'; see also Bonner (1993: 81). The company's bankruptcy is confirmed by a set of accounts for KAS enterprises which is in the author's possession.

[40] *Herald* 11.7.89 'SA Undercover Operation Against Poaching in Doubt'.

[41] *Mail and Guardian* 19.1.96 'Report takes WWF to "Tusk"'. The Kumleben Commission also found that KAS Enterprises was accountable to no one, but that those ultimately responsible for it did not have the intention of intelligence gathering for apartheid South Africa.

number of conservationists and Parks officers publicly deny the existence of corruption, or at least play down its significance. For example, a donor report on funding for rehabilitation of Gonarezhou stated that, because Gonarezhou is so close to the Mozambique border, it was often pretended that most poaching activities emanated from the neighbouring state (Ellenberg et al., 1993: 22). In addition, the successive droughts of the 1980s and early 1990s allowed wildlife authorities to cover up the extent of commercial poaching in the area. In the course of researching this book, it was extremely difficult to find anyone willing to discuss the possibility of poaching by the military. One informant commented that those who had obtained evidence of poaching by the military were mostly dead as a result.[42] There was a strong Renamo, Frente de Libertação de Moçambique (Frelimo) and SADF presence in Gonarezhou in the 1980s and, when military forces are in wild places, poaching occurs either for food rations or because sales of wildlife products provide funding for operations. Reports of poaching by the ZNA in Gonarezhou National Park during the 1980s continued to surface. Despite official denials, the involvement of the military in poaching in the 1980s appears to be widely accepted by conservationists inside and outside Zimbabwe. This is significant because it has an effect on policies to combat poaching and the types of funding made available by donors.

Illegal hunting by internal poachers has had a long history in the southeast lowveld, and especially in Gonarezhou. The strategy of 'freezing' national parks was employed during the liberation war, and afforded Rhodesian security services the opportunity to poach with impunity. One source suggested that the poaching was also linked to an ex-Rhodesian Army syndicate.[43] This is said to have included a man, Ant White, who after the Rhodesian war was co-opted by SADF and was reported to have moved to Beira. He is *persona non grata* in Zimbabwe, but reports suggested he was in and out of the country on a regular basis, because he was protected by the highest levels of government.[44] Indeed, in a recent case in South Africa, a convicted apartheid assassin, Dirk Coetzee, alleged that a South African intelligence unit was responsible for the assassination of the Swedish Prime Minister in 1986, and in his evidence he implicated White as one of the killers.[45] In the mid-1980s reports of large scale poaching in Gonarezhou began to emerge. In 1986, it was reported that 260 elephants and 32 rhinos were illegally killed by poachers, 'who mostly came from Mozambique'.[46] Gonarezhou National Park was the site of Renamo activity and became a frozen zone, closed to

[42] Pers. comm. from anonymous informant.
[43] *Mail and Guardian* 4.10.96 'Palme's Murder Still a Mystery'; see Daly (1982) for further discussion of Rhodesian involvement.
[44] Anonymous interviewee; see also Pearce (1991: 70); EIA (1992: 20).
[45] *Mail and Guardian* 4.10.96 'Palme's Murder Still a Mystery'; *Guardian (UK)* 5.10.96 'World Round Up: South Africa'.
[46] *Sunday Mail* 24.8.86 'Hundreds of Jumbos Poached in Reserve'.

visitors and under military command, in 1987 (it reopened to international visitors in mid 1994). Renamo were blamed for poaching in the park, but EIA research suggested that it was the ZNA with the complicity of wildlife officials (EIA, 1992: 20).

Conservationists outside Zimbabwe, and a few individuals within Zimbabwe, were convinced that members of ZNA had an extensive poaching and smuggling operation in Mozambique and Gonarezhou. A military officer whose nickname was Rex Nango was reported to be at the centre of the poaching and selling of ivory for personal gain in the 1980s.[47] As a frozen zone, there was little communication between the northern and southern portions of the park and, once the ZNA moved into a deal with Renamo incursions in the north of the park, it became a poaching free-for-all. In addition, since the railway to Zimbabwe from Mozambique crossed the border at Sango, when the military declared Sango unsafe, trains carried on to Rutenga, 150 km inside Zimbabwe, where there were no customs controls (Currey and Moore, 1994). The ZNA stated that this decision was based solely on security, but anecdotal evidence suggests that ivory was transported from Rutenga, a place well serviced by roads and so ivory and rhino horn were quickly distributed from there. In fact, two of those interviewed mentioned white-owned trucking companies involved in moving ivory poached by the military.[48]

Since the end of South Africa's policy of destabilisation and the consequent opening of the area for tourism, poaching has subsided in the south eastern lowveld, but the main organisers of the poaching ring that operated in the 1980s have not disappeared and consistently try to find new ways to reopen poaching rings in the area.[49] This is clearly at odds with the stereotype of black African corruption and poverty-stricken poachers so favoured by Western conservation organisations. Complicity from Parks officials was essential to commercial poachers and there were persistent rumours of a poaching faction in the Parks Department, whose alliances in the ruling party were related to the power struggle in the Department.[50] A number of key figures were involved in the poaching in Gonarezhou, besides the military. Bill Taylor, an American dentist, was one of the individuals accused of large-scale commercial poaching, but in 1995 the Parks Department granted him a hunter's licence and, when the news leaked out, the Department said he should never have been given a licence and an investigation would take place. In fact, a Parks spokesman stated that the warden who had first alerted the Department to Taylor's activities was now giving covert support to Taylor's operations.[51] In Gonarezhou, a former warden and a game scout were implicated in

[47] Interview with Stephen T. Bracken, Second Secretary, US Embassy, 21.3.95, Harare.
[48] Anonymous interviewees.
[49] Interview with Clive Stockil, Chair of the Save Valley Conservancy, 13.5.96, Chiredzi.
[50] Anonymous interviewee. See also Chapter Two for further discussion.
[51] *Financial Gazette* 9.2.95 'Parks Grant Notorious Poacher a Hunters Licence'.

poaching. However, rather than being investigated, they were merely moved to other posts in Gokwe and Matetsi safari areas and Chizarira National Park where poaching suddenly increased after their arrival (EIA, 1992: 20; Duffy, 1999: 115).

Critics of Zimbabwe's Parks Department argue that there has been an extensive cover-up of military and official involvement in poaching by conservation authorities and by the government. The difficulties involved in attempting to expose official involvement in poaching were highlighted by the deaths of investigators suspected of having evidence. The most famous case is that of Captain Nleya of the ZNA. He discovered that his own company was involved in poaching rhino and elephant in Gonarezhou in 1988. Reports suggested that he argued with his commanding officer who is said to have pulled a gun on him. In 1989, he reported to Hwange police station that his life was being threatened. That year he was found hanged in his barracks in Hwange and death by suicide was recorded. However, his family kept the case open and later, in 1989, the local magistrate agreed that he was murdered (Thornton and Currey, 1991; EIA, 1992: 20). The Central Intelligence Organisation (CIO) was implicated in the murder. Although Captain Nleya's family maintained that he was murdered and Amnesty International called for the investigation to be reopened, the case petered out. His family complained that they had been threatened by the CIO. Nleya himself told his family that he believed he was being followed by the Army Special Investigations Branch shortly before his death. He was convinced that he was under surveillance because he had threatened to report the poaching and smuggling he witnessed to the highest levels of the armed forces and to the government (Amnesty International, 1992; Duffy, 1999: 115). Clearly, such allegations would have incurred the wrath of his superior officers stationed in Gonarezhou National Park and Mozambique.

Corruption is often cited as the central factor in the continuance of poaching in Zimbabwe. Without corrupt officials within the Parks Department, poachers would not have been able to hunt at such levels and sell on their trophies. Official and military corruption is particularly problematic, because in effect it neutralises the enforcement capabilities of the state. Still, the overriding image of the poacher presented to the public by the Zimbabwean government and by Western conservation organisations is of greedy, poor, black Mozambicans and Zambians, assisted by the corrupt practices of post-independence governments. This is clearly only part of the picture since consistent official denial of the involvement of the military and Parks officials in poaching demonstrates a lack of commitment to combat a certain type of poacher. The inability to tackle poachers in the army and Parks Department has serious implications for Operation Safeguard Our Heritage, which expanded the role of the armed forces stationed in remote areas, thereby affording certain elements in those forces an increased opportunity to poach.

Commercial and subsistence poaching are forms of resistance to certain types of wildlife policy. Such hunting reveals fundamental differences between the state and individuals in the perception of ownership of wild resources. Commercial poaching is related to local subsistence poaching in some ways. A number of the Zambian poachers involved in the rhino wars in the Zambezi valley entered into poaching for economic reasons. While the poachers certainly made large profits from the sale of rhino horns, the amount they received was a fraction of the value of the horn to dealers and smuggling rings. The rhino wars demonstrated the extensive coercive powers of the state, which were employed in full force and with deadly effect in the controversial shoot-to-kill policy. These rhino wars provided an opportunity for international NGOs and donors to raise their public profile. The allocation of funds and equipment to assist Operation Stronghold and Operation Safeguard Our Heritage strengthened the capacity of the Parks Department. It is clear that, without NGO and donor support, shoot-to-kill would not have been operated so efficiently. The war on poachers neatly filled the gap left by the end of the liberation war for military personnel who transferred to the Parks Department, and the strategy was devised and conducted as a new war to protect wildlife.

However, there was a second war being waged on wildlife in the 1980s, which received less attention. The implication of regional armies and Parks officers in ivory poaching and smuggling was subject to a cover-up until the mid-1990s. It was only after the changes in South Africa and the removal of key conservationists in Zimbabwe's Parks Department that these issues came to be debated. The Kumleben Commission in South Africa publicly exposed the extent of SADF involvement in ivory poaching and smuggling. It was clear that poaching was part of the policy of destabilisation devised by the apartheid state. It was intended to drain the front-line states economically by removing valuable ivory and rhino horn and by tying up their armies and Parks Departments in anti-poaching activities. Wildlife products were used to fund destabilisation campaigns, and ivory and rhino horn sales were used to back UNITA and Renamo. Within Zimbabwe, the implication of the ZNA, former Rhodesian Special Forces officers and Parks officers in poaching has cast doubt on the Parks Department's claims that it is an excellent wildlife manager. The consistent denial of these activities has become a significant problem for Zimbabwe in international forums. NGOs, such as EIA, have denounced the Parks Department and its claims to be an effective wildlife protector capable of conducting a controlled and legal trade in wildlife products. The implication of the ZNA and Parks officers was important because the wildlife authorities could not use the same anti-poaching methods against those groups as they had done in the rhino wars, as the coercive powers of the state could not be used against other sections of the state apparatus. While commercial poaching necessitated policies that rely on state-sponsored coercion, the inability of the Parks Department to control poaching in the southern part of the country demonstrates the limitations of coercion. Indeed, such state agencies have assisted various commercial poaching organisations by ignoring or covering up involvement amongst their own armies and, in the case of apartheid South Africa, commercial poaching rings were utilised as part of the regional destabilisation campaign. It was clear that the politics and economics of conservation were far from neutral or non-political, but rather that they intersected with a number of political objectives for apartheid South Africa and were used to devastating effect.

4
Privatising Wildlife Conservation

A major component of Zimbabwe's policy of sustainable utilisation is the concern to find means of conserving wildlife on land outside national parks. This is quite different from a number of other African states, notably Kenya and Tanzania, which have, until recently, focused their efforts on protected areas. In Zimbabwe, the government remains responsible for wildlife conservation within the Parks and Wildlife Estate, but since the 1960s wildlife conservation has been increasingly transferred to the private sector through policies that encourage the devolution of authority and responsibility for wildlife to the landholder, coupled with the definition of wildlife as an economic resource. This trend arose out of fears expressed in the 1960s that wildlife was rapidly disappearing outside the Parks and Wildlife Estate. Wildlife populations within protected areas were becoming increasingly isolated, with the concomitant problems of a smaller genetic pool, established migration routes were severed and some protected areas were being degraded by overpopulation of certain wildlife species. The transfer of conservation to the private sector was also intended to serve two other purposes: to ease pressure on falling budgets for wildlife and to make wildlife less subject to personal interests and power struggles in the government. The policy choice of privatising wildlife conservation is related to ideological underpinnings that view wildlife as a commercial resource to be exploited like any other. The notion of privatised wildlife is politically controversial in terms of domestic politics, given the racial disparity in ownership of productive land. Privatising wildlife is also internationally contentious, because of the commitment to state-managed protected areas amongst conservation non-governmental organisations (NGOs) and the wider environmental movement. Nevertheless, those interest groups involved in privatised wildlife conservation present the policy as a practical management strategy, which, despite being politically contentious, results in enhanced wildlife conservation. This chapter will investigate why wildlife has been gradually moved into the private sector. It will

examine the two aspects of the private sector that are involved in wildlife
– the tourism industry and wildlife ranching. It will highlight the
potential for private-sector involvement in the conservation of endangered species, especially the black rhino, through an analysis of the
lowveld conservancies.

Why involve the private sector?

The debate over whether the state or the private sector is best able to
manage the environment is one that divides environmentalists. The
market and the state can be characterised as two different sets of
management principles. The state is usually preferred as a wildlife
manager, because it is widely perceived as a publicly accountable institution, which can devise management plans and has the capacity to
implement them. In contrast, environmentalists who promote
management by the private sector do so because they argue that individuals seek to promote their own welfare and so protect the resources on
which their welfare depends, and also because economic values for
resources provide the most efficient way of allocating resources.
Environmental management necessarily implies the pursuit of long-term
goals and deferred gains and values. Yet state agencies, which are subject
to changes in government or ruling party policy, may emphasise short-
term goals, such as elections, in the case of democratic states, or
government propensity to acquire personal wealth while in power,
which often work against the interests of wildlife conservation.
Supporters of environmental economics argue that their management
principles promote long-term goals over short-term ones, but the use of
market principles does not necessarily lead to a long-term approach,
since short-term profit maximisation can adversely affect the environment. The principles of environmental economics can be used to
build in values to environmental care that maximise individual welfare,
including assessing the value of goods that are consumed without
passing through the market (firewood, fodder), valuing commercially
harvested products (fish, wildlife) and valuing the indirect values of
ecosystems (watershed protection, preventing soil erosion) (McNeely et
al., 1990). This, in theory, results in parts of the private sector conserving
the resources on which their continued welfare or profits depend
(Barbier, 1987: 101–7; Winpenny, 1991). For example, the continued
survival of wildlife in natural-looking habitats is critical for the tourism
industry and so tour operators have an interest in conserving wildlife in
the long term.

Effective institutional arrangements are essential for the success of
environmental planning. State agencies are often responsible for specific
policy areas, such as agriculture or wildlife, whereas in the private sector
these boundaries may not exist. For example, under the 1975 Parks and

Wildlife Act, communal and commercial farmers in Zimbabwe have been empowered to carry out conservation in their areas. This means that the person or community designated by the state as the landowner or land-holder carries out the functions that might otherwise be the duties of potentially competing ministries. This devolution of authority and responsibility for wildlife is in accordance with the Parks Department policy of sustainable utilisation. The 1975 Parks and Wildlife Act defines the appropriate authority for wildlife management on private land and commercial farms as the owner or occupier of that land (Ministry of Environment and Tourism, 1992: 5). The Parks Department has provided the legislative and regulatory framework within which private sector conservation programmes take place. The role of the Parks Department is to 'facilitate development of a diverse, resilient wildlife industry without prejudice to wildlife conservation' (Ministry of Environment and Tourism, 1992: 5).

To this end, the Department has carried out its own research and collaborated with NGOs to investigate the potential of a private wildlife industry. For example, the Department's *Research Plan* highlights the need to develop an understanding of the economics of the wildlife industry and of the marketing systems necessary to obtain the full value of wildlife (DNWPLM, 1992a: 5). In encouraging the wildlife industry, the Parks Department also fulfils its policy strategy of treating wildlife as a commodity, and wildlife is presented to the private sector as a land use strategy able to compete with cattle and crops.

The financial situation in the Parks Department also provided an important impetus for transferring wildlife to the private sector. Financial problems constitute a significant constraint on effective institutional action and are especially acute in junior ministries, which are unable to attract the large sums they may require to function effectively. Finance is a major problem for the Parks Department and has been since its inception, but it became increasingly clear in the 1980s that conservation budgets were not able to keep pace with spiralling conservation costs. This was especially the case with rhino conservation, which involved very costly anti-poaching programmes and intensive protection policies. The budget allocation for wildlife declined in real terms each year, resulting in poorly-paid Parks officers in comparison with police and army personnel in similar positions. It was clear that the budgets required for this level of protection and conservation were not forthcoming from the government or from external donors (DNPWLM, 1992b: 63–7). The private sector was viewed as a means of resolving budgetary problems. The Parks Department believed that, if the private sector were allowed to use wildlife as an economic resource, it could provide the kind of finance necessary for effective conservation. Economic priorities coincided with political reasons for allowing the private sector to take control of wildlife. It became clear in the 1990s that the power struggle within the Parks Department had the potential to seriously damage the status of wildlife in Zimbabwe. The conservationist faction felt that it was necessary to move wildlife into private hands in order to prevent its overexploitation and possible destruction by corrupt elements within the ruling party.[1]

The private sector

Matthews and Richter argue that the ideological basis of development through tourism growth reveals that tourism is a political process, rather than a private or individualised activity (Matthews and Richter, 1991: 120–3). The renewed emphasis on outward-orientated growth and the rise of neoliberal development strategies have focused attention on tourism as a potential growth sector. The central core of neoliberal development strategies is an emphasis on economic diversification, particularly a commitment to non-traditional exports, such as tourism (Brohman, 1996: 48–52). This approach has also been favoured by the international lending agencies, such as the World Bank and the International Monetary Fund (IMF), and by bilateral donors, which have made loans available in return for reforms that favour market-orientated growth (World Bank, 1994).

Development strategies based on tourism are very much part of the debate about comparative advantage. Advocates of this position argue that each state should concentrate on exporting goods that it is naturally best at producing (Amsden, 1990; Porter, 1990). It is difficult to fit tourism into traditional notions of import and export, because it is an exported good that is consumed by consumers from the North in the countries of the South. However, it is viewed as a foreign exchange earner, in much the same way as traditional exports. Developing countries are considered to have a comparative advantage in tourism, in that they attract tourists from the North, who seek sunshine, beaches and other natural and cultural attractions found in the South. Governments in the South, facing financial problems and an end to secure markets for their goods in former colonial powers, have recognised that tourism can provide an answer to their problems (Lynn, 1992: 371–3). Most governments, regardless of their political ideology, accept the importance of tourism, and so tourism has been an area for significant levels of state intervention, which is unusual for a business that is primarily regarded as a matter for private enterprise (Hall, 1994: 28–32). National tourism policies tend to be geared towards the generation of economic growth and the concept of tourism development is regarded as almost synonymous with economic growth, Westernisation and modernisation for governments. Tourism means employment, balance of payments, regional development and foreign exchange (Harrison, D., 1992b: 8–11;

[1] See *Financial Gazette* 24.3.94 'Department Replete With Corruption'; *Herald* 23.3.94 'Time for Action on Corruption' (letter from 'Public Enquiry Please').

Hall, 1994: 112–20;). Many governments are keen to couch public-sector tourism development in terms of a pro-business rhetoric and policy (Matthews and Richter, 1991: 120–3).

The tourism industry in Zimbabwe relies on wildlife, because visitors come specifically to view animals in their natural habitat. This revenue generating capacity mitigates some of the very high costs of protection and maintenance of national parks and wildlife in developing states (Zimbabwe Trust, 1992). For example, in 1990 the World Tourism Organisation (WTO) estimated that tourism was worth US$62.5 billion for developing states, and argued that many of them were well placed to take advantage of ecotourism, which is the fastest-growing tourism sector (WTO et al.,1992: 1). Developing states have turned to ecotourism as a means of earning foreign exchange, while ensuring that the environment is not degraded. Ecotourism has become an increasingly popular label attached to various forms of alternative tourism. In general, ecotourism involves travel to natural areas with little conventional tourism development, and is travel that concerns itself with flora, fauna, ecosystems and the culture of local people (Boo, 1990: 1–3).

The official tourism policy for Zimbabwe is that of high-cost–low-impact ecotourism development. This can be defined as development that maximises the economic gain from tourism with the minimum negative effect on the environment, meaning that the Zimbabwean government encourages a smaller number of visitors who are willing to pay higher sums for a high quality, personalised service. In this sense, it is the opposite of mass tourism. This ecotourism still relies on the notion that developing states have a comparative advantage in terms of the variety and extent of 'unspoiled' natural environments (Cater, 1994: 69–72). Like many other developing countries, Zimbabwe has been attracted by the notion of ecotourism as a means of developing without jeopardising environmental conservation. The tourist industry is a major revenue earner for Zimbabwe and, in 1993, tourism was the third foreign-exchange earner after agriculture and mining, with droughts in 1994 and 1995, tourism became the second largest after mining. Government statistics demonstrate that over the period 1991 to 1993 the number of visitors rose from 696,659 to 971,539, with a corresponding increase in expenditure on travel, tours and accommodation by those visitors (Central Statistical Office, 1993). In 1995, tourism reached a new peak, with a record one million visitors.[2] The tourist industry is dependent on attractions such as Victoria Falls, and, more importantly, on wildlife. The Zimbabwe Tourism Development Corporation (ZTDC) noted that tourists were motivated to visit by notions of exotic appeal, uniqueness or a sense of 'real Africa', the friendliness of the people, good infrastructure and security, a cultural or historical dimension to the trip and value for money, with the most popular trips being game safaris (ZTDC, 1993;

[2] *Financial Gazette* 6.7.95 'Tourists Hit One Million Mark'.

Dann, 1996a). The tourist industry is a diverse one, which also means that the tourist industry creates ancillary economic activities, such as transport companies, accommodation providers, street venders and souvenir/curio sellers in tourist areas.

Tourist activities all rely on the presence of wildlife in areas where tour operators can gain access to them, such as national parks. In southern Africa, the regional income from tourism is skewed, with Zimbabwe, Botswana and South Africa receiving the greatest share of a globally expanding market (Cumming and Bond, 1991: 26; WTO et al., 1992: 15–17). Zimbabwe is well placed to take advantage of the growing ecotourism market, with small-scale wildlife safaris on private lands. The Parks Department *Research Plan* makes it clear that tourism is perceived as the key to the economic survival of the Parks and Wildlife Estate and to justifying wildlife utilisation as a form of land use (DNPWLM, 1992a: 43). For example, it is possible to place tourist viewing values on different species, such valuation having been most developed in Kenya, where it was estimated that in 1989 the viewing value of elephants in Kenya was approximately US$20–30 million (Leader-Williams, 1994a: 63). This was because the Western tourists that make up the bulk of Kenya's tourist market placed seeing an elephant high on their list of priorities in a wildlife safari. In Zimbabwe the consumptive values of wildlife have also been developed through sport hunting and this means the economic values of wildlife are enhanced, because they can be sold for viewing and then sold again for hunting (Wildlife Society of Zimbabwe, 1993).

Since wildlife is central to the economic well-being of the tourism industry, many operators consider long term planning for conservation as essential. For example, one of the major hotel chains, the Zimbabwe Sun Group, recognised that 'the long term success of this industry is totally dependent upon the harmonious interface between tourism development and the environment in which that development takes place' (Smith, 1994: 23). For the Zimbabwe Sun Group it is clear that the environment will only be conserved when it is seen as a significant economic resource, and it claims that its own development projects in Katete at Kariba Dam and Mahenye on the borders of Gonarezhou National Park were intended to provide major and tangible economic benefits for the company and for the people in the locality, so that protection of the environment assumed greater importance (Smith, 1994: 24–5). It is clear that tour operators do not simply argue in favour of more effective conservation. Tour operators assist conservation in very direct ways by providing an important source of funding and capital goods for some areas of wildlife conservation. Safari and tour operators often make significant contributions to the areas in which they operate, particularly in the form of road maintenance and providing diesel for vehicles (Cunliffe, 1994). For example, a private sector organisation, the Zimbabwe Council for Tourism, donated Z$2 million to the Parks Department in 1993 because it was concerned at the lack of funds in the Department. The tourist industry acts as a critical support base for state sector involvement in wildlife conservation. The industry support stems directly from its recognition that continued survival depends on the conservation of wildlife in the national parks and safari areas. Of course it also provides private tour operators with good publicity within Zimbabwe and with overseas visitors.

There are a number of problems associated with tourist development. Zimbabwe's policy of high-cost–low-impact tourism is intended to avoid some of the difficulties associated with mass tourism as found in Kenya, such as overcrowding of visitors in national parks, unplanned and uncontrolled building of tourist facilities and resorts and the adverse publicity that such tourist development often attracts. The problem of visitor carrying capacity is faced by all states attempting to develop their tourist industry. In Zimbabwe, that capacity is relatively low and so advertising concentrates on attracting a small number of visitors because the industry and the facilities in national parks and other scenic areas would be unable to cope with the demands of mass tourism.

However, in attracting only wealthy tourists from overseas, the tourist industry and government have to attempt to reconcile possible conflicts between local people and tourists. The presence of wealthy tourists raises potential cultural and class problems, such as the question of whether separate facilities should be provided for overseas visitors in national parks (Cumming, 1990a). The Parks Department's policy is to have a two tier system of charges for entry into national parks, one for overseas visitors and one for Zimbabwean citizens. The Parks Department argues that Zimbabweans contribute to the maintenance of national parks through taxation – for example, in 1995 Zimbabweans contributed Z$44.8 million in tax revenue to the Parks Department – and so it is argued that they are entitled to a discount when compared with overseas visitors, who only pay into the system when they are visiting.[3] This, in effect, means that local people are not outpriced by foreign visitors and have a real and usable right of access to enjoy their own natural heritage at a reasonable rate. In accordance with this, the private sector tour operators and hotel operators offer a range of facilities that are open to locals and overseas visitors. The two tier rates and range of facilities assist in mitigating some of the problems attendant on tourism, especially the criticism of tourism as a process that privileges wealthy overseas visitors over local people (Ap, 1992; Munt, 1994a,b).

One final problem with reliance on tourism to fund conservation is that tourism can be fickle, since a steady flow of tourists relies heavily on image (Dann, 1996a). National parks, in particular, continue to rely on the romantic imagery of wilderness and wildlife to attract visitors. This reliance on the imagery of an untouched wilderness contains problems for tourism, because the experience could be spoiled by large crowds, so

---

[3] *Herald* 11.2.95 'Explain Discounts in Parks Entry Fees'; *Herald* 4.9.93 'Entry Fees for Locals at Resorts'; *Daily Gazette* 4.9.93 'Locals Now Pay to Visit National Parks'.

protected areas have to restrict access either by cost, by relative inaccessibility or by quotas. Reliance on tourism to pay for conservation is also problematic because areas of great conservation importance, such as swamps, are not necessarily major tourist attractions (WTO et al., 1992: 6–17). On a wider scale, globalisation of Western culture is perpetuated through the tourist industry, which carries with it implicit representations of Africa as a gigantic outdoor zoo (Zimbabwe Trust, DNPWLM and the Campfire Association, 1990). In addition, adverse publicity, civil strife and changes in Northern holidaying tastes can all dramatically affect the economic viability of a tourist industry (Norton, 1994; Bloom, 1996; Hall and O'Sullivan, 1996; Sonmez, 1998; Sonmez and Graefe, 1998).

However, it is likely that African wildlife will continue to be attractive to Western visitors and that the value of wildlife tourism on the continent as a whole will increase. It will be important for individual countries to retain their popularity as tourist destinations, but Zimbabwe faces difficulties because it is not well known as a tourist destination. For example, South Africa has become an increasing threat to Zimbabwe's tourist industry because the Western public is familiar with it and, since the end of apartheid, has begun to perceive it as a respectable destination. In addition, Tarirai T. Musonza of the ZTDC claimed that many tourists believe Victoria Falls is in South Africa, especially since special flight options to the Falls have been added to South African tour itineraries.[4] There are a number of problems attendant on reliance on the tourist industry as a promoter and guardian of wildlife conservation. The main difficulty is that, if tourism ceases to be a major foreign exchange earner, then the industry's underpinning of state conservation efforts would be jeopardised, and so it is difficult for such an industry to guarantee a long-term and sustained commitment to wildlife protection.

A second form of private-sector involvement in wildlife conservation is the rapidly growing wildlife ranching industry. It has become a significant contributor to wildlife conservation in national and international terms. The development of wildlife ranching in Zimbabwe is presented by the Parks Department as a great conservation success. The models used for encouraging commercial farmers to move into the wildlife sector were almost directly translated into the Communal Areas Management Programme for Indigenous Resources (Campfire) programme for communal farmers. Private wildlife ranches have become an increasingly important sector for conservation and are perceived as a real and workable alternative to state responsibility for wildlife.

Wildlife ranching has its origins in the 1960s, when legislative changes encouraged commercial farmers to conserve wildlife, but it was the 1975

---

[4] Interview with Tarirai T. Musonza, Research and Planning Officer, ZTDC, 23.3.95, Harare; *Sunday Mail* (South Africa) 26.3.95 'Meeting the Challenge of the Tourism Industry'; *Financial Gazette* 6.7.95 'Tourists Hit the One Million Mark'.

Parks and Wildlife Act that proved to be the major catalyst in private sector wildlife conservation. Under the Act, private landholders received appropriate authority status, which gave them the right to dispose of wildlife as they wished. This change in legislation resulted in a fundamental change in attitude for many ranchers, who proceeded to conserve wildlife on their land (Child, 1988: 182–336). Critics argued that such legislative changes would result in the total disappearance of wildlife, but in fact the opposite occurred. In 1994, it was estimated that 75% of private ranches derived some or all of their income from wildlife-based ventures and that the area allocated to wildlife was expanding at 6% per annum during the 1990s (Martin, 1994a: 7).

Wildlife utilisation has proved attractive to private ranchers for a number of reasons. First, the relative returns from cattle and wildlife changed from the 1960s onwards, as the prices for beef and tobacco declined, and wildlife utilisation proved a more attractive alternative. In addition, patterns of consumption in the world beef market have changed. There has been a swing away from consumption in Europe and North America, which is matched by greater demand from the Pacific rim region. Zimbabwe exports beef under the Lome Conventions to the European Union, but is given no such concessions in the growing Asian markets (Cumming and Bond, 1991: 19–31). Consequently, a chance to move away from beef production was especially welcome in the case of the arid areas characterised by low or erratic rainfall and poor soils, where the costs of producing wildlife were much lower than the costs of producing cattle in an environment to which they were not naturally suited. Research by the World Wildlife Fund (WWF) into the wildlife ranching sector also revealed that wildlife farming is less ecologically damaging than commercial cattle ranching (Cumming and Bond, 1991: 93–124). The transition from cattle farming to wildlife farming was a smooth one. Most ranchers had some wildlife populations left on their cattle ranches and simply allowed them to build up slowly over time, during which ranchers could still derive income from cattle until wildlife populations were sufficiently large to utilise sustainably (Cumming and Bond, 1991: 87–93). It was only in South Africa, Zimbabwe and Namibia that ranches large enough to support viable wildlife populations were found (Cumming and Bond, 1991: 19–31).

Despite the lack of detailed data on the wildlife industry in Zimbabwe, it is clear that a switch to wildlife farming had significant benefits, rapidly proving that it was a profitable form of land use. Ranchers in the lowveld of Zimbabwe turned to game ranching because the returns were greater than from cattle. Ranchers derive income from sport hunting (which is the major income earner), selling surplus wildlife stock to other ranchers or Campfire areas, wildlife based tourism and, to a lesser extent, cropping game for meat sales (especially impala).

Sport hunting by overseas clients is the major revenue earner for the private game ranching industry. In the short term, sport hunting provides the greatest returns with minimal investment in facilities for visitors. For example, in Zimbabwe, the sport hunting industry grew from a value of US$195,000 in 1984 to US$13 million in 1993, and the number of hunts rose from 25 to 1,300 over the same period.[5] This growth was assisted by the special reservations on sport-hunted ivory and other wildlife trophies that Zimbabwe lodged with the Convention on International Trade in Endangered Species (CITES), as a result of which sport hunters can export their trophies from Zimbabwe to their home country with the relevant documentation (Craig and Gibson, 1993). There are a number of difficulties with reliance on sport hunting for revenue generation. One significant threat is the development of the animal welfare movement in the North. The continued survival of the American sport-hunting market is critical to Zimbabwe's ranchers. In 1989 approximately 63 per cent of all overseas sport hunters came from the USA, but Zimbabwe's continued commitment to sport hunting has been criticised by animal rights and animal welfare groups, especially in Britain and the USA.[6] The promotion of sport hunting also has the potential to discourage Western tourists who are interested in phototourism, since, as a result of the expansion in the animal rights movement in the North, tourists can be encouraged to boycott states that do not adhere to strict wildlife preservation. In addition, the reservation of areas for sport hunting by overseas hunters raises class dimensions and, to take account of this, Zimbabwean hunters are provided with lower-cost hunting on Zimbabwe Hunters Association (ZHA) land and in state-owned safari areas to prevent locals being outpriced by foreign hunters.[7] Finally, the abuse of hunting permits is also a problem, since there is always a possibility that a hunter will shoot one trophy animal and shortly after see another with a better trophy. In some cases, corrupt safari operators might be tempted to allow the hunter to shoot the second trophy and sell the original trophy for personal profit (Bonner, 1993).

Another form of revenue generation is to sell surplus wildlife stocks to other ranches and Campfire areas. As the game ranching sector has become more developed and sophisticated, wildlife ranchers realise that newcomers to the industry need to restock their ranches with the wildlife they had once shot out. In the 1990s, a system of wildlife auctions has become firmly embedded in the wildlife ranching industry. The first auction of wildlife was held in 1986 to restock ranches denuded of wildlife.[8] In 1993 record prices were paid for sable (Z$24,000 per head) and elephants averaged a price of Z$13,000.[9]

[5] *Financial Gazette* 12.1.95 'Sport Hunting Quotas Cut'.
[6] Interview with Keith Madders, Director of Zimbabwe Trust (UK), 12.10.94, Epsom, Surrey; and interview with David Cumming, Director of WWF-Zimbabwe, 22.2.95, Harare.
[7] Interview with Charl Grobbelaar, Chief Executive of Zimbabwe Hunters Association, 12.2.95, Harare.
[8] *Ziana* (Zimbabwe) 10.1.86 'Game Sale Expected to Net Over $100,000'.
[9] *Herald* 29.7.93 'High Prices Paid for Sable at Wildlife Auction'; *Financial Gazette* 14.7.94 'Demand for Wildlife Exports Rises'; *Herald* 19.8.94 'Wildlife Farmers Report Rise in Animal Export Demand'.

There is one possible difficulty with such auctions. It is the problem of the freeloader in a classic 'tragedy of the commons' situation (Hardin, 1968: 1243–8). Wildlife ranches are normally grouped together into a larger wildlife area, so that the removal of internal fencing allows free movement of wildlife. Rowan Martin pointed out that there was a danger that an individual rancher might sell more than just surplus stock if it was a bad financial year.[10] Such an action might prompt other ranchers to leave the scheme. However, the freeloader problem is less likely to occur in the groupings of ranches, because the system of sets of ranches has created a group access situation, rather than the classic open access situation, which is often blamed for environmental degradation (Abel and Blaikie, 1990). It is in all the ranchers' interests to ensure that wildlife populations are maintained at a viable level for hunting, cropping, sales or phototourism, because it is necessary for the ranchers to conserve wildlife in order to ensure the survival of their business. In this sense, Zimbabwe's wildlife conservation policy creates economic interests in favour of wildlife conservation.

Generation of revenue from wildlife viewing is closely related to the wider tourist industry in Zimbabwe, as discussed earlier. The number of wildlife ranches that rely on photographic safaris and special-interest tours (bird-watching, botany, foot safaris and horse safaris) has grown in recent years. The initial dependence on sport hunting as an immediate source of revenue was slowly replaced by non-consumptive tourism. These developments have been strongly supported by the Wildlife Producers Association (WPA) and the Commercial Farmers Union (CFU). For example, the WPA formed the Zimbabwe Safari Farms Co-op Ltd, the purpose of which is to promote and market tourism on farms and ranches that can offer tailor-made safaris in unique tracts of bush and luxury accommodation (WPA and CFU, 1992: 2). The increased interest in wildlife tourism reflects the profitability of that kind of enterprise, and indicates a recognition that phototourism has the potential to be more profitable in the long term (Weiler and Hall, 1992).

Consumptive utilisation of wildlife for production of skins, hides, horns and meat produces relatively less revenue, but impala cropping has remained an important part of the wildlife ranching industry. For example, in 1989 the WPA estimated that 20,000 impalas were cropped on game ranches per annum.[11] Cropping of wildlife is the least economically productive method of utilising wildlife. In cropping, wildlife is sold only once for its products, while wildlife used for photographic tourism can be sold multiple times.

It is clear that the shift into wildlife production on commercial farmlands has occurred for a number of important reasons. There are political

---

[10] Interview with Rowan Martin, Assistant Director: Research, Department of National Parks and Wildlife Management (DNPWLM), 29.5.95, Harare.
[11] *Herald* 26.1.89 'Game Meat Expansion'.

reasons, such as the need to avoid seizures of commercial land by the government under the 1992 Land Acquisition Act, and the recognition in the Parks Department that to leave wildlife in the state sector might result in its mismanagement. There are also economic reasons, such as the decline in the terms of trade for agricultural products, the relative profitability of wildlife utilisation over cattle and the global growth in sport hunting and in viewing and photographic tourism.

The lowveld conservancies: privatising black rhino conservation

The establishment of large-scale private conservancies has been a significant development for wildlife conservation in Zimbabwe. Private conservancies are considered by the Parks Department, some donors and NGOs and by a number of commercial farmers to be the great hope for the future of wildlife conservation in southern Africa. Supporters of the private conservancies promote them as models of effective conservation, coupled with some local development. Clive Stockil suggested that economic sustainability, ecological sustainability and a recognition of the importance of the socio-economic needs of surrounding communities were the motivating factors for the transfer to wildlife production in the lowveld.[12]

The development of conservancies grew directly out of the provisions of the 1975 Parks and Wildlife Act, which conferred authority over wildlife on the landowner, and without this legislative framework the concept of a private conservancy could not exist. The positive experience of conservancies and large game ranches in South Africa was also a significant factor in fostering interest in the conservancy concept in Zimbabwe. The term conservancy can be applied to any number of properties joined into a single complex in order to ensure better management, conservation and utilisation of some or all of the natural resources within that area (Price Waterhouse, 1994: 17). The ranchers in the area were involved in game ranching and mixed systems of cattle and wildlife before the conservancy was established. The ranchers came together to cooperate in wildlife management, because they realised they could reduce the costs of anti-poaching, fencing and wildlife restocking. Conservancies are run on the principles that the members are jointly responsible for recurring management costs, that there is no internal fencing and that management is based on sound scientific principles. The development of a legally binding constitution for members of a conservancy was a major breakthrough, because, until then, the private ranches that make up the Save Valley and Bubiana Conservancies had

[12] Interview with Clive Stockil, Chair of the Save Valley Conservancy, 13.5.96, Chiredzi.

operated together on an ad hoc basis (Leader-Williams, 1994b: 4–8). Those directly involved with the conservancy perceive the constitution as the key, since it ensured that members were legally bound to the conservancy concept and obliged to adhere to certain management principles. In terms of achieving national policy objectives in conserving endangered species, the conservancies are perceived as *the* hope for black rhino conservation.

The droughts of the late 1980s and early 1990s highlighted the fragility of the lowveld's ecology. It was clear that decades of exploitative cattle ranching had caused widespread environmental damage. One of the main motivators for a switch from cattle to wildlife was a local rancher, Clive Stockil. He was concerned at the damage done by cattle, such as that numerous grasses had been grazed out of the area and productivity had declined over time.[13] It was necessary to repair the damage, particularly around the Save River Basin (Price Waterhouse, 1994: 26). The ranchers involved in the conservancy had realised that dependence on a single species, cattle, meant that production was easily affected by drought, and the transfer to wildlife was intended to result in greater resilience in times of drought. In addition, internal fencing was removed from the conservancy to allow more freedom of movement for wildlife, because the conservancy was also based on the notion that the environment and wildlife in the lowveld could be better managed on a larger and more extensive scale (Price Waterhouse, 1994: 26). One of the main justifications for the establishment of a large private park was that it would undoubtedly be of ecological benefit in the fragile lowveld. Promoters of the conservancy have consistently argued that wildlife production is more naturally suited to arid regions with poor soils than are crops or cattle.

The survival of conservancies is dependent on their financial viability. As private enterprises, they must make a profit from wildlife utilisation in order to ensure wildlife conservation. The economic rationale of conservancies relates directly to the national government policy of obtaining the maximum value for wildlife. In the feasibility study carried out by Price Waterhouse, a theoretical case study of a wildlife enterprise versus a cattle ranch was presented. Price Waterhouse concluded that under the conditions found in the Save and Bubiana conservancies, cattle would generate Z$445,171 (gross), whilst wildlife had the potential to yield nearly double that, at Z$800,480 (gross) per annum (Price Waterhouse, 1994: 60–4). In 1994, the members of the Save Valley Conservancy took the final bold step of removing the last cattle from the area in order to concentrate on wildlife utilisation.[14] Sport hunting and

---

[13] Interview with Clive Stockil, Chair of the Save Valley Conservancy, 13.5.96, Chiredzi; De La Harpe (1994). *Daily Telegraph* (UK) 24.10.94 'Farmers Turn from Cattle to Big Game in Zimbabwe'.

[14] *Daily Telegraph* 24.10.94 'Farmers Turn from Cattle to Big Game in Zimbabwe'.

tourism are the main forms of wildlife utilisation. As the economic value of wildlife viewing is enhanced by the presence of the 'big five', cattle had to be removed in order to introduce buffalo (Price Waterhouse, 1994).

It was clear, from the experience with national parks, that it was necessary to take account of local interests if the conservancy was to succeed. It was important that the conservancy did not have the definite fenced and patrolled borders that had led to the increasingly criticised model of protected areas perceived as having a hard edge, and so local communities were encouraged to feel part of the Save Valley Conservancy. Eighty-four per cent of the Save Valley Conservancy boundary is bordered by Communal Lands and indigenous resettlement areas, and the lowveld contains some of the most densely populated Communal Areas in Zimbabwe, which are entirely situated in the fragile Natural Region V (Metcalfe, 1996b: 4). The *Black Rhino Conservation Strategy* states that areas surrounding private conservancies must be friendly territories, because the risk of local involvement in commercial poaching of rhino had to be minimised, and because if a rhino escaped from the conservancy it would have to be located and captured quickly and easily (DNPWLM, 1992c: 8). Consequently, in order to ensure the viability of the conservancy, it was necessary to ensure that people in the local area benefited directly from the venture. The Save Valley Conservancy has begun a programme that is modelled on Campfire in the surrounding Communal Areas. In order to give local people a stake in conserving wildlife in the conservancy, a trust was established. The trust manages relations between the conservancy members and local communities, and revenues intended for use in communities (Metcalfe, 1996b: 4). Save Valley Conservancy members envisage that the trust will raise Z$65 million from donors, which will then be used to purchase wildlife. The trust will then be a shareholder in Save Valley Wildlife Services Ltd.– it will own 64 per cent of the shares and the landowners will own 36 per cent – and it is hoped that, through tourism and future wildlife sales, the trust will generate Z$5 million per annum in perpetuity for local communities to spend on development projects (Metcalfe, 1996b: 3–4; Save Valley Conservancy Trust, [1996?]). As with the Campfire programme, giving local people a stake in conservation is also part of a broad anti-poaching strategy. Benefits can be provided in terms of employment either directly, in the conservancy, or indirectly, in Communal Lands in industries that grow out of the conservancy. For example, local people were employed in building tourist facilities and have been involved in ongoing operational activities (Price Waterhouse, 1994: 8–11).

One of the key justifications for private conservancies is that they can provide more effective protection from poachers than state-run protected areas. In conservancies, poaching has been tackled through traditional anti-poaching patrols and cooperation with local communities. Intelligence gathering and regular patrolling have been a feature of conservancies and the Save Valley Conservancy, for example, has one full-time ranger and three game scouts, who are responsible for tracking black rhino and responding to individual poaching incidents on member ranches. In addition, each ranch has its own game scouts to monitor wildlife, providing ground coverage of one scout per 25 square kilometres, which is greater than in national parks. Scouts have been issued with radios, which have contributed to cooperation and intelligence gathering within the conservancy.[15] Dehorning rhinos was also carried out in the hope that it would act as a deterrent. Clive Stockil argued that poachers may be deterred if they are aware of the high level of ground coverage, coupled with the knowledge that they will only obtain a stump, raising the risks for the poacher while reducing the potential benefits.[16] Detection is a central factor in anti-poaching in the conservancies. The conservancy members recognised that traditional military-style patrols and coercive tactics would not provide the best means of preventing illegal hunting. As a result, a strategy was devised that was based on inducements rather than coercion. The fact that surrounding communities are shareholders in the wildlife venture is intended to make local people more resistant to poachers and more likely to inform on them. This is supported by a system of rewards that has been publicised in Communal Lands. The conservancy offers higher rewards for information than the amount poachers can offer for information on the whereabouts of rhinos and all those involved in the conservancy have been encouraged to report suspected poachers.[17] Clive Stockil pointed to the conservancy's excellent record on poaching, with no major incidents since 1991, as evidence of the efficacy of a strategy based on inducements rather than coercion.[18]

Black rhino conservation has had a central role to play in the development and continued financial viability of the conservancies. The black rhino crisis provided a catalyst for the formation of the conservancies, although rhino protection was not their fundamental purpose. In the 1970s a small group of white rhinos were introduced to the Humani Ranch in the Save valley area, and this proved to be the nucleus of the conservancy established in 1991. In 1986–7 more black rhinos were introduced from the troubled Zambezi valley. It was their tendency to stray on to neighbouring properties that convinced the ranchers of the need for joint management and prompted the formation of a conservancy. Similarly, the Bubiana Conservancy was set up with the primary motivation of black rhino conservation (Price Waterhouse, 1994: 31–7). The Save Valley Conservancy holds 42 white rhinos and four black rhinos, on

[15] Interview with Graham Connear, Chief Conservator, Save Valley Conservancy, 13.5.96, Chiredzi.

[16] Interview with Clive Stockil, Chair of the Save Valley Conservancy, 13.5.96, Chiredzi.

[17] Interview with Raoul Du Toit, DNPWLM/WWF, 7.5.96, Harare.

[18] Interview with Clive Stockil, Chair of the Save Valley Conservancy, 13.5.96, Chiredzi.

the basis that they are part of a commercial undertaking; that is, the rhino must be a sufficient attraction to tourists to justify their presence and the cost of their conservation.[19] Indeed, the tourism potential of the conservancies has been linked to the development of the south eastern lowveld as a world superpark. It was envisaged that the conservancy would eventually be linked to Gonarezhou National Park, Kruger National Park, an area in Mozambique and a protected area in Swaziland (Price Waterhouse, 1994: 19–23; Peace Parks Foundation [1997?]).

The poaching wars that centred around the black rhino in the Zambezi valley provided the impetus for the Parks Department's decision to move rhino into private conservancies (Leader-Williams, 1994b: 4–8). It was clear that the state-owned national parks were not adequately protecting the rhino, and it was hoped that private conservancies would have the necessary finance and staff to ensure proper protection. The *Zimbabwe Black Rhino Conservation Strategy* stated that the conservancies should be at least 50 kilometres from the nearest international border to reduce the external poaching threat (DNPWLM, 1992c: 8). It was envisaged that the private conservancies could provide adequate protection from poachers. NGOs, such as Sebakwe Black Rhino Trust, WWF, the Beit Trust and Tusk, have been involved in fund-raising to equip employees in the conservancies with radios, vehicles and weaponry to carry out anti-poaching patrols (Leader-Williams, 1994b). The strategy of moving rhinos on to private land was also perceived as an insurance policy against extinction. It removed the all-eggs-in-one-basket approach of having all rhinos on state land, where they were clearly at risk from poachers, and it also eased the budgetary pressures that arose in the Parks Department from the poaching wars (Price Waterhouse, 1994: 19–23). The black rhino has a special status in Zimbabwe, which means it remains the property of the state wherever it is found, because it is so endangered. This has raised the curious situation of a resource that is state-owned being conserved and paid for by the private sector. Dependence on private conservancies means that rhinos are placed in a situation of relative wealth at minimal cost to the state, which technically still owns them (Leader-Williams, 1994b: 14). The Parks Department guidelines state that private conservancies that are able to hold at least 100 rhino will be given translocated rhinos from the Zambezi valley. As a result, viable founder breeding herds of at least 40 rhinos should be established in private conservancies (DNPWLM, 1992c: 8). The rates of increase in the rhino population in the lowveld conservancies are very encouraging: 9 per cent per annum, with an increase of 3 per cent for the Midlands Conservancy (Leader-Williams, 1994b: 8–11).

---

[19] Interview with Charl Grobbelaar, Chief Executive of Zimbabwe Hunters Association, 12.2.95, Harare. The ZHA is involved in running the Conservancy and in raising funds for it; a number of Conservancy members are also ZHA members. Since sport hunting of rhino is not permitted, rhino are kept as a tourist attraction.

The difficulties encountered by the new private conservancies indicate the wider issues of reliance on the private sector to conserve natural heritage. The problems faced by conservancies indicate important political issues facing the wildlife ranching sector. The policy environment, which has so far encouraged private sector involvement, has been criticised for the way it empowers some interest groups over others. The conservancies are still very much at the experimental stage of their development. Nevertheless, a number of difficulties arose in the first few years of their operation that indicate their relationship to national political and economic problems.

First, poaching remained a problem for the conservancies. The Price Waterhouse feasibility study noted that subsistence poaching was rife and that there was some commercial poaching (Price Waterhouse, 1994: 35). The possible link up with Gonarezhou and the conservancy's proximity to the area known as crooks' corner, where Zimbabwe, Mozambique and South Africa meet, has raised concerns about the security of rhinos. Anti-poaching has also proved more problematic than expected by the conservancy members. The question of private armies involved in using coercion to protect privately held resources has arisen. As a result, scouts in private conservancies are not indemnified in the same way as Parks scouts, and so they cannot fire on suspected poachers with the confidence that they will not be charged with unlawful killing. In addition, lack of equipment for anti-poaching operations significantly impaired the conservancies' abilities to protect the rhinos. This was a particular difficulty for Bubiana Conservancy. The lack of adequate anti-poaching operations is especially problematic, because one of the main reasons for the Parks Department support for conservancies was that they were better able to protect rhino from poachers.

It was perceived by the members of the conservancy that the main donors, WWF and Beit Trust, would provide adequate communications equipment and weaponry for anti-poaching patrols, but this was not the case. WWF's policy was not to supply weaponry or grants for weaponry, highlighting differences between the expectations of aid recipients and those of aid donors. The misunderstanding arose from the fact that Beit Trust was responsible for grants for weaponry and this became inextricably linked with support from WWF in the minds of the Bubiana Conservancy members. The willingness to place blame on WWF perhaps also stemmed from the highly politicised appointment of the principal ecologist, Raoul Du Toit. Du Toit was responsible for rhino conservation in the lowveld conservancies, was paid for by WWF, but was ultimately accountable to the Parks Department. His time was largely consumed in Harare which frustrated conservancy members and donors (Leader-Williams, 1994b: 14–19). Originally, it was intended that the conservancy meet the costs of day to day anti-poaching and general management. For their part, WWF and Beit Trust agreed to employ a suitably-qualified ecologist to provide technical assistance in setting up wildlife monitoring systems. However, in the course of finding a suitable candidate, the conservancy members pressured the donors to provide someone who could coordinate anti-poaching efforts and also provide more finance for operational budgets. WWF was concerned that this constituted a change of policy and so did not agree to the request. This case highlights the gap that often exists between the commitments that an aid recipient perceives to have been agreed and the actual donor conditions. After months of wrangling, the weapons were removed by the Parks Department, after fears were raised about private armies. This caused friction between the conservancies and the Department.[20]

The wider political context of conservancies, especially their implications for land redistribution, heavily influenced the ways that conservancies were justified by their supporters to the Zimbabwean public and internationally. The land question has been the central defining political issue for conservancies, because they constitute a visible example of the retention of large areas of land by white ranching interests. In 1996, the Minister for Environment and Tourism pledged to curb the expansion of private conservancies even in areas that lie outside regions that are suited to subsistence and commercial-crop farming.[21] The ranchers involved in the Save Valley Conservancy perceive the Land Acquisition Act as a threat to their plans. After the expiry of the Lancaster House agreement, which protected land owned by the settler community, the Land Acquisition Act was drawn up in 1992, under the terms of which the government can compulsorily purchase land that it defines as derelict or underutilised. Since the introduction of the Land Acquisition Act, commercial ranchers have been keen to ensure that their land is economically productive. Consequently, ranchers have turned to the luxury safari industry to justify their continued ownership of large tracts of land. This has been the case in the lowveld, where cattle ranching became less profitable due to drought and long term environmental damage done by cattle.[22] Cattle ranchers who had previously failed to manage their estates efficiently found transferring to the wildlife sector a welcome opportunity. Amongst members of the government, there is a perception that land under wildlife is underused and the Land Acquisition Act presented a real political threat, because it could be used to close conservancies and open the land for resettlement or crop and meat production.

The legal framework for conservancies has remained unclear. The conservancies developed rapidly, while the government has lagged behind in terms of providing a regulatory framework.[23] The Land Acquisition Act constitutes a lesser threat to the conservancies in very

---

[20] Interview with Clive Stockil, Chair of the Save Valley Conservancy, 13.5.96, Chiredzi.

[21] *Hansard* 14.2.96, 'Written Answers to Questions', 3968–71, The Minister of Environment and Tourism, Mr Chimutengwende.

[22] *Herald* 25.11.91 'Game Ranches Sprout'.

[23] *Sunday Gazette* 17.12.95 'Private Conservancies Boom Worries Government, Farmers'.

arid areas, such as Save and Bubiana, but has already proved to be a major issue in areas where crop and beef production are more profitable. The issue of food security has been used to justify curbs on private conservancies. The Minister for Environment and Tourism, Chen Chimutengwende, stated that the development of conservancies had to be closely controlled in order to prevent them threatening food security in Zimbabwe.[24] In addition, the Deputy Minister of Lands and Water Resources, Cain Mathema, pledged that he would monitor the expansion of game farming to ensure that it did not encroach on croplands, fearing that the switch to game farming would jeopardise Zimbabwe's ability to produce enough food and export crops.[25] However, the fears for food security have not been used as an argument against tobacco and cut flower production, both of which are grown on prime croplands and are not food crops.[26]

The transfer of rhino from the Midlands Conservancy to the lowveld conservancies by the Parks Department highlights this issue of which areas are suited to wildlife farming. The Midlands Conservancy is situated in the area south of Harare and extends towards Kwekwe. It predates the lowveld conservancies and is situated on land that is better suited to agriculture and cattle. The less favourable rate of rhino population growth and the relatively high numbers of rhino deaths in the Midlands Conservancy were blamed on unsuitable habitat (Leader-Williams, 1994b: 19). It was decided that rhinos which did not reach their full breeding capacity would be removed from private ranches and given to the Bubiana Conservancy in order to establish a viable breeding herd in a suitable habitat (DNPWLM, 1992b: 52). The removal of rhinos to the lowveld conservancies was perceived by the Midlands Conservancy as favouritism on the part of the Parks Department (Leader-Williams, 1994b: 19). The transfer also underlined that fact that rhino remained a state-owned resource, regardless of the powers conferred on landowners by the 1975 Parks and Wildlife Act. The main concern for the Midlands Conservancy was that, without black rhinos as an attraction, visitors would rather go to the lowveld conservancies, where they could be guaranteed a sighting. In turn, their profitability might be jeopardised and so they would be unable to escape the terms of the Land Acquisition Act.

The private conservancies and the development of wildlife ranching have been criticised because they perpetuate the existing unequal distribution of land and economic power. As a result, another seemingly innovative area of wildlife conservation, which was presented as a practical management plan, intersected with highly controversial issues in the

---

[24] *Hansard* 14.2.96, 'Written Answers to Questions', 3968–71, The Minister of Environment and Tourism, Mr Chimutengwende; *Herald* 4.11.95 'Moves to Curtail Private Ownership of Wildlife'.

[25] *Sunday Mail* 11.6.95 'Crop Land Safe From Game Farmers'.

[26] Interview with Ivan Bond, Resource Economist, WWF-Zimbabwe, 6.5.96, Harare.

domestic political context. This was because it supported the existing allocation of land, power and control over lucrative resources. The political and historical factors that resulted in white commercial farmers being the main beneficiaries of conservancies have alarmed some elements in the government. The vast majority of commercial ranchers are white settlers, who benefited from the colonial policies of reserving the best land for European Areas (Palmer, 1977; Naldi, 1993). The resettlement issue has been used as a tool to deny the importance of the wildlife farming sector on the grounds that such empty lands should be made available for resettlement from crowded Communal Lands.[27] The conservationist faction in the Parks Department has been accused of restocking the ranches owned by their friends with animals that belong to the state and the people of Zimbabwe.[28] Consequently, the well-stocked private conservancies were considered by some government members to be a potential threat to the national parks. However, as stated earlier, one of the reasons that the Parks Department decided to move wildlife into the private sector was to attempt to wrest control over wildlife from corrupt government officials. The ranching sector has been criticised as a hangover from the colonial period, which has not been remedied by the post-independence government, which promised a comprehensive resettlement programme. Consequently, very few black Zimbabweans are involved in the commercial wildlife ranching sector.[29] The conservancies have recognised that failure to indigenise will ultimately threaten their long-term survival, and so they have tried to attract black entrepreneurs into the ventures.[30] However, wildlife has continued to be perceived by the wider Zimbabwean society as a white-controlled area of the economy. In addition, indigenisation has been regarded with suspicion amongst the conservation community for being a smokescreen for corrupt acquisition of resources.

Finally, one of the main issues facing the development of conservation on private land is an economic one. Blue-green strategies for conservation, which rely on providing wildlife with an economic value, are vulnerable to changing economic priorities. If the black rhino is defined as part of the national heritage, its future conservation should be guaranteed. Such a level of commitment is more likely to come from a state institution than from a private commercial venture. Mike Hitschmann, of the Wildlife Society, stirred up controversy by suggesting that the government was correct to curb private conservancies, because it was wrong to leave all rhinos in private hands, where a few individuals would profit from them.[31] It was financial considerations and not conser-

---

[27] *Sunday Mail* 11.6.95 'Crop Land Safe From Game Farmers'.

[28] *Sunday Mail* 27.11.94 'Private Game Parks Threaten Department'.

[29] *Herald* 3.3.94 'Big Hurdles Face Our Would be Farmers of Wildlife'.

[30] Interview with Tom Taylor, Chief Executive of the Save Valley Conservancy, 10.5.96, Harare; and interview with Raoul Du Toit, DNPWLM/WWF, 7.5.96 Harare.

[31] *Herald* 15.7.95 'Don't Confine Rhinos to Private Land Says Expert'.

vation objectives that first attracted cattle ranchers to wildlife farming; consequently, financial considerations could also lead them away from wildlife conservation and back to cattle or some other profitable form of land use. As the economic viability of the conservancies is dependent on tourists interested in wildlife viewing, they could be threatened by changes in the tourist flow. Regional or national strife, changes in holidaying tastes, increases in air fares or a switch back to higher levels of red meat consumption in the Western world could all upset the economics of wildlife ranching.

In conclusion, the involvement of the private sector in conservation is highly developed in Zimbabwe. The tourism industry provides vital funds for the Parks Department and other conservation activities in national parks. This commitment is motivated by the recognition that without wildlife the tourism industry would be less profitable. In addition, the rapid growth of the wildlife ranching sector has been an important element in the overall national policy for conservation. Wildlife ranching is in accordance with the theoretical underpinnings of Zimbabwe's conservation policies – that landowners are best equipped to be environmental managers. The moves towards wildlife by the private sector have been supported by the conservationist wing of the Parks Department for a number of economic and political reasons. It is not motivated by a belief in the capabilities of the private sector *per se*, but more by the notion that the private sector will at least be better equipped than state agencies to manage wildlife. Amongst the ranchers, the motivation for the switch to wildlife ranching was economic, since the failure of cattle revenues to keep pace with the profits to be had from wildlife farming provided a major impetus to farmers to change over. The large private conservancies constitute an essential element in conservation. Their role in black rhino conservation has been particularly important for the Parks Department. As with the rest of the private sector's involvement, the conservancies have eased budgetary pressures in the Parks Department.

However, the economic motivations for the switch to wildlife are problematic. The difficulties arise if wildlife becomes unprofitable, for whatever reason. In this respect, state responsibility for national parks and wildlife is a necessary element of an overall national conservation strategy, but state responsibility can also fall foul of economic problems, particularly a lack of adequate finance to manage the environment and protect endangered species. For the Parks Department, there were political reasons for the moves to private sector conservation, as well as economic ones. The Parks Department recognised that, in the face of continuing budget cuts, it could not afford to conserve wildlife throughout its range, or even just in national parks. The devolution of authority to Campfire areas and the private landowners can be viewed partly as a cost-saving exercise. In addition, political motives stemmed from the need to wrest control of wildlife away from the state and give it to private individuals and groups. This was intended to prevent the corrupt use of wildlife by certain elements in the Parks Department, the Ministry of Environment and Tourism and their allies in the government. The fact that blue-green approaches tend to reinforce the existing distribution of land, money and political influence has led to criticisms. This has especially been the case in Zimbabwe, because this conservation policy intersects with the racially unequal distribution of land and resources and has raised sensitive political and social issues, which affected the domestic political acceptability of a powerful and largely white-owned wildlife ranching sector. These obstacles within the domestic political environment are critically important to the future success of conservancies. Failure to adequately negotiate the legitimate calls for an end to the racial disparity in land ownership would rightly deliver a fatal blow to private ownership of wildlife.

5

Community Participation in Campfire

as Political Legitimation

In recent years, the buzz-words in conservation and rural development have centred around ideas of participatory management of wildlife. The Communal Areas Management Programme for Indigenous Resources (Campfire) in Zimbabwe has attracted national and international attention as a programme that is at the forefront of an innovative and workable approach to negotiating the conflict between people and wildlife. As such, Campfire is the key to political legitimation on the international stage for the controversial approach of sustainable utilisation. In its policy, the Zimbabwean Parks Department has ensured that the parks-versus-people relationship is not the central focus of wildlife policy, and instead the dilemma has been recast to present it as one not of conflict, but of inclusion and coexistence. Wildlife conservation and rural development have traditionally been considered as conflicting. Conservation requires existing areas of land for wildlife to be maintained, if not increased, whereas, in contrast, rural development is thought to necessarily mean more land for crops, livestock or industry. This conflict between conservation and rural development is most sharply demonstrated by the national parks systems of sub-Saharan Africa. National parks resulted in local people being moved from their land, excluded from the new parks and denied access to the wildlife and grazing areas that they once enjoyed (Ranger, 1989). The failings of the exclusionary approach were demonstrated through continued poaching and utilisation of national parks resources by local people. In a number of African states, national parks became paper parks, which exist on maps and in tourist brochures, but in fact their boundaries are not respected by local people.

The Parks Department in Zimbabwe devised a system that relied not on exclusion and penalties, but on inclusion and incentives to conserve wildlife. The Campfire policy is widely acclaimed by conservationists inside and outside Zimbabwe. Campfire has been promoted as the answer to the potentially conflicting processes of conservation and rural development, which, in Campfire, are viewed as mutually reinforcing and interdependent. In accordance with the stated national policy of sustainable utilisation of natural resources, Campfire treats wildlife as a resource and operates on the principle that if wildlife has an economic value for rural people, they have a reason to conserve it, and revenues derived from it will be spent on rural development. The scheme has a number of powerful supporters, including rural development agencies, local non-governmental organisations (NGOs), donors and wildlife conservationists. Campfire has provided a focal point for arguments between conservationists in the international arena over utilisation of wildlife for economic gain and has proved Zimbabwe's most effective argument against strict preservationism. However, its reliance on sport hunting and the ivory issue have been used by its opponents to criticise the programme as unethical. This chapter will explore the concepts and mechanisms of Campfire. It will investigate the history of the programme and its objectives, including the empowerment of rural people, the revenue distribution process and the creation of incentives to remove conflict with wildlife. It will also assess the role of sport hunting and the ivory trade in generating revenue for Campfire and it will analyse the importance of NGOs in implementing and supporting the scheme. Finally, it will investigate the role of the state in community partici-pation.

Local interests and the objectives of Campfire

There is a perception that Campfire began as an idea hatched by a group of white liberals in the Parks Department. The innovative nature of Campfire meant that the group Murphree referred to as the 'khaki shorts brigade' of wildlife enthusiasts found themselves at the forefront of debates about rural land reform and communal tenure (Murphree, 1995). Despite the perception that Campfire was not aimed at empowering indigenous Zimbabweans, it was quickly embraced by a number of rural districts. Once independence came, the 1975 Parks and Wildlife Act looked discriminatory and very colonial, and in 1982 the Parks and Wildlife Act was amended to allow District Councils in the Communal Areas to be designated as an appropriate authority to manage wildlife. This legislative change allowed the concept of Campfire to be further developed during the 1980s, but the first Campfire areas were only estab-lished in 1989, in Guruve and Nyaminyami.

In Campfire areas, three sets of ownership rights and interest groups intersect, namely local rights, private interests and state interests. Part of its importance, in terms of a political legitimation of the utilisation approach, is that Campfire is directly concerned with producing local interests that are in favour of wildlife conservation. Martin's working document of 1986 set out the four objectives on which Campfire was founded. These are to obtain voluntary participation in programmes for long-term natural resource management, to introduce a system of group ownership, to provide appropriate institutions to manage resources for the benefit of the community and to provide technical and financial assistance for communities to enable them to realise their goals (Martin, 1986).

In line with ideas of sustainable development, Campfire aims to remove the conflict between environment and development in order to ensure long-term satisfaction of basic human needs. Sustainable utilisation draws on the concept of a safe minimum standard of offtake, which will ensure that the resource is not depleted. It is also informed by the idea that poverty is a major cause and effect of environmental degradation. Willie Nduku stated that Campfire could be considered as two-pronged, since for the government it is a conservation tool to prevent poaching and wildlife habitat destruction, while for rural people it means wildlife represents development, rather than signifying backwardness.[1]

Supporters of Campfire point to the expansion of land under wildlife conservation and this has been used as the cornerstone of the arguments to gain domestic and international political and economic acceptance for a policy that allocates large areas of land to wildlife conservation in the face of repeated claims for land redistribution. Proponents of Campfire point out that, while the Parks and Wildlife Estate covers just over 12 per cent of Zimbabwe's land area, when Campfire areas and ranches under wildlife are added, the area of land under wildlife is 33 per cent and continuously growing.[2] Unsurprisingly, the first Communal Lands to choose Campfire were those already relatively rich in wildlife. However, Campfire areas have begun to buy wildlife to restock their lands in order to benefit from increased hunting quotas or tourism opportunities.[3] Such a rise in wildlife numbers and increases in the area of land available for wildlife have been used by supporters of Campfire as evidence of its success. In the face of potentially hostile social and economic forces, wildlife utilisation has been deemed to be the best means of reconciling conflict between environment and development.

The question of how communal farmers can be persuaded to conserve wildlife on their lands is central. Landholders require incentives to turn their attention to wildlife and away from crops and livestock. Consequently, a discussion of the motivations of rural people has become a focal point of Campfire. The most commonly used models of the peasantry in rural development literature are the rational actor and moral

---

[1] Interview with Willie Nduku, Director, Department of National Parks and Wildlife Management (DNPWLM), 7.7.95, Harare.

[2] Interview with Brian Child, head of the Campfire Coordination Unit, DNPWLM, 16.5.95, Harare.

[3] *Herald* 26.11.89 'Centenarians Don't Want to Lag Behind'; *Sunday Gazette* 26.3.95 'Communal Sanctuary for Rhino'.

economy models (Scott, 1976; Popkin, 1979). The way that communal farmers are defined and perceived to operate draws on the wider debates about rural development. The central focus of Campfire is that people will care for their environment when they perceive that the benefits exceed the costs of doing so. Caring for the environment and Campfire require investment, consideration of the future and a willingness to accept delayed returns.

In a break with pre-independence attitudes, Campfire began from the assumption that rural people are rational decision-makers. Campfire's acceptance of rural people as rational actors that live in a moral economy of sharing resources and revenue reflects an increasing interest in harnessing indigenous knowledge for conservation and rural development. Stephen Kasere, of the Campfire Association, suggests that the ready acceptance of Campfire by some communities lies in the fact that it constitutes a reinvigoration of traditional environmental beliefs and not because of the meat and money that it brings. For example, in the precolonial period wildlife was regarded as common property and subject to cultural controls, and breach of codes by killing wildlife was punishable in mundane and spiritual ways. In addition, certain groups were not allowed to kill totem animals (in Matabeleland, the surnames Ndlovu, Dube and Nyathi refer to elephant, zebra and buffalo, respectively) (Kasere, 1995a: 6–9). The end result of such environmental beliefs was the protection of species. The conviction that rural people are the ones who are best able to manage the environment in which they live has only recently come to be accepted in a few conservation circles. This is not to suggest that rural life is unchanging or backward. However, Martin also points out that indigenous people should not be perceived as being allowed to use wildlife because they have no other means of generating an income, as such a view is patronising and reveals an unspoken text that it is all right for indigenous peoples to use wildlife. Instead, wildlife utilisation should be accommodated and accepted by the dominant cultural norm (Martin, 1994b).

Property rights and land tenure have been two major concerns in debates on rural development issues. Rights and land tenure have proved to be important factors in the development of Campfire. One of the key assumptions in the environmental movement is that, if resources are treated as common property, people's inherent disposition to act in a selfish way means that those resources will be degraded, a process more commonly referred to as the 'tragedy of the commons' (Hardin, 1968). This notion is contradicted by Campfire, because it does not operate on the open-access system that environmentalists and conservationists criticise as responsible for environmental degradation. Instead, it constitutes a group-access approach, but Campfire in practice is still struggling to achieve the ideal of strong communities with clearly defined rights. Wildlife is considered a community asset, but communities are organisationally complex (Child, B., 1995a; Matose, 1997: 77).

Murphree suggests that, in Zimbabwe, Campfire was not based on any single theoretical model of common property regimes. Instead, a pragmatic approach was taken, which drew on debates about common property, politics, economics and wildlife conservation. This approach was described as adaptive management, meaning that what worked was continued and what did not work was discontinued.[4] The experience of commercial farms informed the decision to confer strong property rights on rural communities, but it was recognised that land units in the Communal Areas were not analogous to those in commercial farming areas (Murphree, 1995). However, Willie Nduku pointed out that, in Communal Areas, the land is not owned by anyone, but is utilised by the people on it. In the same way, wildlife is not owned by the community but is utilised by them, and so wildlife is a resource held in a group system, which is governed by usufruct rights.[5] Problems with defining ownership of wildlife also meant that it was difficult to decide who or what constituted an appropriate authority over wildlife. This was why the District Council, rather than the individual farmer, was established as the focal point of the initial institutional structure of Campfire in Communal Lands. The notion of boundaries surrounding common resources is extremely important. Common-resource management is problematic where the resource and its appropriators are poorly bounded (Metcalfe, 1992a: 36). Zimbabwe is fortunate in having clearly defined ward and district boundaries around communities in rural areas, and this was built on by introducing a new system of group ownership with strictly defined rights of access to wildlife for the communities resident in that area (Murombedzi, 1992: 13).

Campfire relies on decentralisation of ownership from the state to local communities. This, in turn, requires communities to feel empowered to be able to take control of resources and manage them. This is not an easy task after decades of colonial styles of governance, which were top-down and based on enforcement and coercion rather than incentive. Nevertheless, Campfire supporters have been keen to stress the need to engender bottom-up development. The process of empowerment depends on some definition of who and what constitutes the community.

The stated aim of the Parks Department is to extricate itself from any management role in Campfire. It is tempting to view Campfire as a money-saving exercise, and certainly the Parks Department was aware that it could not afford to protect all wildlife in all areas without cooperation from local people. Consequently, wildlife management and ownership were devolved to the community by investing new powers in the Rural District Council. It was hoped that eventually devolution

---

[4] Interview with Professor Marshall Murphree, CASS, University of Zimbabwe, 18.7.95, Harare.

[5] Interview with Willie Nduku, Director, DNPWLM, 7.7.95, Harare.

would be extended to wards and finally villages, but as yet there are no legal mechanisms to allow this.[6]

However, those involved in Campfire were aware that communities are heterogeneous bodies. Campfire does not necessarily include all the communities in the district. In some areas, newcomers and those arriving on resettlement schemes have been excluded by the rest of the community from benefiting from Campfire revenues.[7] District Councils face challenges to their authority from within the areas they represent. In general, the two main interest groups in campfire areas groups are characterised as traditional authorities, such as chiefs, and arms of the state, such as the District Councils. Although these are very generalised terms, which carry inherent assumptions about the nature of chiefs and District Councils, they are definitions that are widely used by those involved in Campfire. Traditional leaders and authorities in rural areas often do not coincide with the area under a District Council, or they operate in direct contravention of each other's interests. This is especially the case in allocation of land. In theory, the District Council and central government authorities deal with land allocation in the national resettlement scheme, but in practice it is often the local chiefs who allocate land (Dzingirai, 1994; Child B., 1995a). This led to serious difficulties, because the two authorities made decisions that directly impinged on each other. Chiefs in some areas received payment for allowing illegal settlement in their areas. The illegal settlers viewed themselves as part of the community, because they had paid the chief and were allocated land by him (Dzingirai, 1994: 1–3). Some new settlers were attracted to Campfire areas after hearing reports of the revenue being generated by wildlife and because tsetse eradication campaigns have opened up new areas for human settlement, such as Omay, which were previously deemed unsuitable for heavy settlement (Murombedzi, 1992). The difficulty was that the original settlers viewed the newcomers as outsiders to the community, and denied their claims to Campfire revenue. In Omay, such illegal resettlement led to the area being overcrowded, and land that was put aside for wildlife in 1990 became home to settlers and their cattle. Inevitably, this reduced the effectiveness of Campfire, with land for wildlife reduced and the revenues from wildlife having to be distributed among a higher number of households, resulting in some hostility towards Campfire from the local community.[8]

However, it is difficult to remove squatters, because they have often been given permission by the local chiefs, who can have complicated political reasons for inviting settlers into the area. Traditional authorities

[6] Interview with Brian Child, head of the Campfire Coordination Unit, DNPWLM, 16.5.95, Harare.

[7] Interview with Keith Madders, Director Zimbabwe Trust (UK), 12.10.94, Epsom, Surrey.

[8] *Herald* 14.2.95 'Rocky Desert Looms in Omay', Kanyati; interview with David Cumming, Director, WWF-Zimbabwe, 2.2.95, Harare; interview with Keith Madders, Director, Zimbabwe Trust (UK), 12.10.94, Epsom, Surrey.

can view Campfire authorities as a threat, because Campfire can work against their interests as cattle barons, since they could be denied grazing areas in favour of wildlife (Madzudzo, 1995a,b). The agencies involved in implementing Campfire thought that it was essential that the community defined itself, as people within wards and districts have diverse interests groups, with differing capabilities to work with the project. Differing capabilities inevitably result in disproportionate benefits within the community, and these differences had to be reconciled in some way for Campfire to succeed. Internal community conflict indicates the importance of ensuring that differing interests in wildlife conservation are negotiated and, if possible, made compatible.

Community interests can also be divided along gendered lines. In general, rural development projects use the household as the unit of analysis, and the household has commonly been defined as having a male head. However, in Zimbabwe's Communal Lands, numerous households are headed by women and, as the main producers and providers of food, women are dependent on the environment every day, but the particular role of women and their ability to take advantage of the opportunities that a development project can offer have been overlooked. Participation by women in Campfire has continued at a lower rate than male involvement. In a study of Masoka village in the Zambezi valley, Nabane found that women were under-represented on the Campfire committees in their locality. A number of explanations were offered, including that women were more likely to say that people would not elect a woman, but only male respondents claimed that women lacked the confidence to serve on committees. Women and men agreed that a husband's unwillingness to allow a woman to travel and serve on committees was stifling participation. Some argued that women should not sit on committees with men while their husbands were absent, while others suggested that women's lack of skills in literacy and English was a hindrance (Nabane, 1994: 4–14).

Geographical splits have caused problems for institutions that claim to represent the community. For example, in Chapoto Ward, Guruve District, the wildlife committee members were all from the east bank of the Mwanzamatanda River and yet the tradition of wildlife utilisation and the greatest difficulties with the control of problem animals were on the underdeveloped west bank (Buchan, 1989a: 25). Similarly, conflicts arose between the traditionally wealthy people and the poorer people in the community. The very wealthy cattle barons in rural areas were not in favour of Campfire. Keith Madders of the pro-Campfire Zimbabwe Trust (Zimtrust) suggested that this conflict arose simply out of the fact that Campfire would put their grazing areas under wildlife, and the revenue from the wildlife was to benefit the whole community, while the cattle barons would lose their source of personal wealth.[9] The community is

---

[9] Interview with Keith Madders, Director, Zimbabwe Trust (UK), 12.10.94, Epsom, Surrey; see also Madzudzo (1995a, b).

not necessarily conceptualised as a homogeneous unit in Campfire and, as with many revenue-generating schemes, divisions within the community have arisen or been sharpened as a result of Campfire. Murombedzi argues that it is important that Campfire should not founder on the assumption that communities can easily provide collective management on their own (Murombedzi, 1990: 11). As far as the implementing agencies were concerned, it was crucial to leave the definition of the community to people in Communal Lands and to allow these communities to take their own decisions.

Empowerment is an essential process to enable communities to take their own decisions. The Minister of Environment and Tourism stated that Campfire's success depended on empowering people to manage their own resources.[10] It was necessary to convince rural people that wildlife legally belonged to them and that they would not be punished for using what was theirs. The initial stages of Campfire, then, required a programme of raising awareness to encourage people to recognise their ownership rights and to make rules to govern access to and use of wildlife. This has been the ultimate aim of Campfire – an institutional structure in which communities can carry out their own management of natural resources and maximise the sustainable returns (Martin, 1986). Supporters of Campfire argued that it required democratic structures, where people could discuss their interests and needs in relation to wildlife management.

Empowerment was certainly achieved in some cases. For example, in Hurungwe District, the people of Ward 7 in Nyamakate voted for a refund of their money, which was spent on a grinding mill without their permission by a councillor, the grinding mill was repossessed and the community received individual cash payments instead.[11] Similarly, villagers in Matebeleland North and South began to express their dissatisfaction with the mechanisms of Campfire. They argued that they were not receiving the full value for their wildlife, because they did not play a proper role in marketing wildlife products.[12] However, the process of empowerment has not been achieved with ease, and the 1994 Rukuni Commission, which investigated land-tenure issues in Zimbabwe, clearly stated that the experience of Campfire demonstrated that there was a need for greater empowerment of communities over conservation of their environment (Government of Zimbabwe, 1994b: 29). The differing interest groups in each Campfire area have meant that the policy has been more successful in some areas than in others. However, the supporters of Campfire within Zimbabwe use its very public and visible positive impacts to argue that the broader policy of sustainable utilisation has been a great success.

[10] Speech by H. Murerwa, Minister of Environment and Tourism, to the Fourth Annual General Meeting of the Campfire Association, 16.12.94, Harare; interview with Keith Madders, Director, Zimbabwe Trust (UK), 12.10.94, Epsom, Surrey.
[11] *Hurungwe District Campfire Newsletter* no. 4, October 1993.
[12] *Herald* 5.5.93 'Villagers Express Fears Over Campfire Benefits'.

One of the main objectives of Campfire is that it should reduce conflict between people and wildlife. Commercial and subsistence poaching in Communal Lands were symptoms of the adversarial relationship between animals and people. Supporters of Campfire claim that, in general, Campfire has reduced the amount of subsistence poaching in rural areas. It is clear that the motives for and nature of poaching in Communal Areas differ considerably from commercial poaching for ivory and rhino horn. Campfire areas containing species that are attractive to commercial poachers have been targeted by organised poaching. For example, Nyaminyami Wildlife Management Trust was forced to request semi-automatic weapons from central government to protect rhinos and elephants that remained in their area. It was a thorny issue, because valuable wildlife had to be protected from poachers who were 'outsiders', while wildlife managers had to be sure that potentially counterproductive and heavy-handed tactics were not used against locals (Taylor, 1993: 12).

Subsistence poaching has also been problematic. One safari operator, who has a hunting concession in a Campfire area, commented that it would be impossible to stop all subsistence poaching and that it was inevitable that it would continue at a low level.[13] Those involved in implementing Campfire accept that there are areas, such as Dande, where poaching has continued at a relatively high level, despite Campfire. Nevertheless, Brian Child, of the Campfire Coordination Unit, claimed that, overall, the numbers of wildlife were increasing in Communal Lands which indicated a drop in poaching and an increase in effective conservation measures.[14] Certainly, the rates of subsistence poaching by local people in Campfire areas have been reduced. This is because people have changed their attitude towards wildlife. However, there are continuing difficulties with subsistence poaching. The survival of wildlife is in the interests of rural communities, because it provides significant benefits. In effect, the increased economic value of wildlife has assisted in creating community-based anti-poaching interests.

Sustainable utilisation of wildlife for economic gain is one of the means of reducing conflict between wildlife and people. For rural people, wildlife has costs and benefits. The costs range from direct costs, such as crop losses and loss of human life, to indirect costs, such as shorter school days for rural children, because they must travel in daylight to avoid elephants (Kangwana, 1994: 41–5). There are numerous reports of wildlife destroying crops, killing livestock and occasionally people. For example, a lion, nicknamed Maswerasei, which had continually terrorised villagers in Omay Communal Lands, was offered in the hunting quota for 1993.[15] Compensation from Campfire revenues is

---

[13] Interview Mark, Zindele Safaris, 29.5.95, Harare.
[14] Interview with Brian Child, head of the Campfire Coordination Unit, DNPWLM, 16.5.95, Harare.
[15] *Daily Gazette* 8.11.93 'Maswerasei's Former Victims Benefit From Hunting Proceeds'; *Herald* 21.3.95 'Baboons Destroy Crops'; *Herald* 11.5.95 'Marauding Lions Prey on Villagers Cattle'.

intended to offset the costs of living with wildlife. It is difficult to judge the level at which people will accept cash dividends in return for these losses. In areas where the revenue generated by wildlife use is relatively low, communities do not accept this as compensation for loss of human life or livestock.

Apart from providing economic incentives to protect wildlife, Campfire authorities have used various methods, including electric fencing, to reduce human–wildlife conflicts. Elephants have been responsible for destroying crops and attacking people. In Nyaminyami, in 1990, over 60 per cent of complaints about problem animals were associated with elephants, 20 per cent concerned lions and 16 per cent related to buffaloes (Jansen, 1990: 11). The conflict was reduced by using what the Parks Department terms problem animal control (PAC) and the related sport hunting of elephants (occasionally, the two are combined). If an elephant repeatedly raids crops, it can be shot in PAC operations by commercial safari hunters or Parks Department officials contracted by the local District Council.[16] More recently, chasing the elephants rather than killing them has been encouraged, so that the hunting value of such animals is not lost. Commercial hunting operators were also encouraged to have their clients shoot problem elephants (Taylor, 1993). In many ways, PAC by Parks officials works against the interests of safari operators and communities, because it means that elephants are hunted without any revenue (or profit) being generated.

Electric fencing has been used extensively to reduce the conflict between people and wildlife in Campfire areas. Hoare and Mackie argue that the role of fencing as an attitude barrier is as important as its function as a boundary to problem animals (Hoare and Mackie, 1993: 9). Fencing has been the subject of some controversy. A great deal of donor funding and Campfire revenue has been spent on electric fencing to keep wildlife and land for crops and livestock separate. This prevents wildlife from destroying crops and livestock and, if Zimbabwe exports beef to the European Union (EU), veterinary fencing is necessary to prevent the spread of foot-and-mouth disease, because wildlife (especially buffaloes) is believed to be a vector.[17] Electric fences have been welcomed in some areas as a means of preventing wildlife from entering crop and livestock land and, in areas with few employment opportunities, it has provided temporary work (Buchan, 1989b: 20–4).

However, as with all rural developments, fencing holds benefits for some and costs for others. Veterinary fencing programmes, which were not part of Campfire, became confused with Campfire fencing, when, for example, the EU funded a fencing project in Hurungwe District and the local community assumed it was part of Campfire. The fence was erected

---

[16] *Hurungwe District Campfire Newsletter* no. 4, Oct. 1993; *Mudzi District Campfire Newsletter* no. 3, Oct. 1993.

[17] Interview with Artur Runge-Metzger, EU Delegation to Zimbabwe, 7.3.95, Harare.

without proper consultation and numerous households were moved to accommodate it, and this provided an opportunity for a local politician who was against Campfire to use this episode to assist him in a campaign to remove the local MP, who was pro-Campfire.[18] Some local politicians claimed that Campfire was responsible for evicting people, so that land can be used for wildlife in order to further their own election chances.[19] This is a message that is readily accepted by people in rural areas, who have been moved in the past to make way for wildlife or white-owned commercial farms. Nevertheless, it has been noted that Campfire fencing has inevitably resulted in some people being moved and, although some locals are employed in fencing, it is generally undertaken by professional companies, such as Farm Management Africa (Murombedzi, 1992). Fences keep people and cattle out of wildlife areas, rather than wildlife away from people. This works in favour of safari operator interests, which may be opposed to local communities killing wildlife for food or in self-defence because it reduces their hunting quota. To critics of Campfire, fencing merely repeats the pre-Campfire situation, where people were denied access to wildlife. This preserves the colonial right of access to wildlife for the settler community or wealthy overseas hunters and denies it to black Zimbabweans, who live with the costs of wildlife every day. Neumann (1997) argues that the new fashion for participation and benefit sharing in conservation replicates the coercive conservation strategies previously used, and that it assists in an expansion of state control over rural areas (see also Hill, 1996).

The way that revenues are distributed within Campfire areas is a fundamental factor in the success or failure of the programme. It is crucial that communities see the link between the revenue they receive and the wildlife in their area. The problem that Campfire faces is how to achieve this, since it is not the revenue *per se* that is defined as the critical factor, but rather the way in which it is distributed. It was deemed important that a situation did not develop where villagers were having cash passed down to them by Campfire authorities. Rather, revenue should be generated and distributed by the community, with the minimum involvement of the state. The revenue distribution process is the key to Campfire's role, as evidence of the efficacy and social justice of sustainable utilisation in a developing economy. The argument that wildlife revenues are directly passed to the community, which then decides how the money is spent, is routinely used in debates about the ethical acceptance of utilisation in the developing world. The Parks Department drew up draft guidelines for District Councils to follow when disbursing revenue from Campfire schemes. The guidelines set out five principles, indicating that the majority of the revenue should be returned

---

[18] Interview with Stephen Kasere, Deputy Director, Campfire Association, 14.2.95, Harare; *Hurungwe District Campfire Newsletter* no. 6, February 1994.
[19] *Hurungwe District Campfire Newsletter* no. 4, Oct. 1993.

to the producer community, the choice of how to spend the revenue must be made by the community, the process of distribution must be transparent, checks and balances must be built into the distribution process to make it accountable and the community must be small enough to meet regularly in a forum in which all members can participate (ideally, it should not exceed 200 households). It was anticipated that this revenue distribution process would result in a culture of self-sufficiency, rather than the dependency created by top-down processes (Child B., 1995b: 5). For example, in some communities, the link between wildlife and revenue has been made by contract between the community and the safari operator. In Mahenye, the local safari operator placed Z$50,000 on a table for each of the three elephants shot and smaller amounts for other animals, right down to Z$20 for individual baboons, and then each household received their cash payment in turn and signed for it (Child B., 1995b: 22).

Controversy arose over the community's right to decide how to spend the wildlife revenues. The conflicts at a national level over how Campfire revenues should be spent partly reflected debates within communities. There is a conflict between investing in development projects and individual cash payments. Implementing agencies, such as the Parks Department and the Campfire Association, argue that communities must be allowed to decide on individual cash payments. A critical factor in whether a community decides to invest in community projects or vote for individual dividends is the perceived trustworthiness of the District Council. If rural people believed that the District Council would misuse funds, they tended to vote for cash payments; while this was not a vote for individual payments as such, it was a vote against corruption in local government.[20] There have been cases of misappropriation of Campfire funds. For example, in Nyaodza, Hurungwe District, there were complaints that after the community voted to use their money for schools, the schools had not received all the money they had been promised.[21] Nevertheless, criticisms were levelled at decisions in favour of individual cash dividends. In some areas, the cash dividends from wildlife were spent on beer, because there was no other outlet for money and because some dividends were too small to be used for investment in the individual household.[22] As a result, politicians and critics of Campfire have suggested that individual payments are a waste of money. For example, the Governor of Mashonaland West, Ignatius Chombo, suggested that Campfire was not benefiting rural people, because they shared the money out instead of investing the revenue in development

[20] Hurungwe Rural District Council: Report on Disbursement of Campfire Community Funds in the District, 4.6.94.
[21] Wildlife Coordinators Report for the Month 15 March to 15 April 1994, Hurungwe Rural District Council.
[22] Wildlife Coordinators Report for the Month 15 March to 15 April 1994, Hurungwe Rural District Council.

projects. He publicly expressed his belief that the money should be spent on roads, schools or clinics.[23]

Despite these contests over whether individual payments are appropriate, rural people have chosen to invest in infrastructural projects. In 1993, Guruve District Council disbursed Z$600,000 accrued from hunting projects in six wards in Dande and Guruve. The community decided that the revenue was to be spent on the construction of classroom books, housing for teachers and the improvement of sanitation in the villages.[24] In addition, in 1994, for the first time, it was suggested that, if rural schools built libraries and laboratories with Campfire revenue, the Minister of Education and Culture should agree to let them hold sixth-form classes.[25] Without such facilities in rural areas, pupils have to travel to towns or become boarders, which is too costly for the vast majority of rural people. Where villagers have deemed it appropriate, they have spent Campfire revenues on development projects and, in accordance with the aims of Campfire, revenues have also been spent on protecting the resource that is the source of income. It is significant that, in areas where poaching was the norm for local people, game guards have now been trained to carry out anti-poaching and intelligence gathering (Child, G., 1995).

One of the criticisms of Campfire is that it was initially conceived of as a conservation project by the Parks Department and that it is only dressed up in development language to make it more acceptable to rural people and other (primarily external) interest groups.[26] Campfire has proved to be the most attractive and ethically acceptable face of sustainable utilisation and it is difficult for critics to argue against the moral weight of assisting development in poverty stricken rural communities. Supporters deny that Campfire has been given the veneer of a rural development policy to make sustainable utilisation and sport hunting more palatable to the outside world. However, it is clear that Campfire sorts out a number of difficult issues for Zimbabwe and for external interests. For Zimbabwe, it allows environment and development to be presented as mutually reinforcing processes. It also provides the government with a strong argument for allowing continued wildlife utilisation schemes, for reopening a legal ivory trade to increase rural revenues and for continued sport hunting to benefit rural development. For external interests, such as wildlife NGOs and donors, Campfire is presented as socially aware conservation, and external interests can justify support for sport hunting and wildlife utilisation on the grounds that it is helping the poorest of the poor in a developing country.[27] These different emphases do not necessarily threaten the

---

[23] *Herald* 6.10.94 'Campfire Scheme Not Benefiting People Says Governor'.

[24] *Daily Gazette* 4.6.93 'Guruve Distributes Campfire Proceeds'.

[25] *Herald* 31.10.94 'Wildlife as a Vehicle for Development'.

[26] Interview with Eric Feron, Meat Production Adviser for Nyaminyami Wildlife Management Trust and Institut Français des Recherches Africaines (IFRA), 28.2.95, Harare.

[27] See *Financial Gazette* 27.10.94 'Human Expansion Difficult to Justify?'.

future or success of Campfire, but rather amount to negotiable differences between the interest groups that support the same conservation and development programme. This is important in terms of the Parks Department's use of Campfire as the central legitimating argument for sustainable utilisation, since it would be more difficult to get utilisation accepted on the international and national stage if it appeared that the very groups Campfire was meant to benefit were unsympathetic or critical.

Private and state interests: the implementing agencies

Private and state interests are involved in revenue generation, institution building, funding and representation of rural interests in Campfire. The multiple interests involved mean that there are considerable layers of support for Campfire, which assist the Parks Department in using the policy as the key to their political justification of sustainable utilisation, in order to deflect criticism in the national and international arenas. The private sector is deeply involved in Campfire. It is represented by safari operator interests and by NGOs. These agencies have claims over the ownership and use of wildlife. Safari operators play a central role in revenue generation for Campfire areas, and as such they exercise considerable control over how wildlife resources are utilised. NGOs provide support mechanisms for Campfire, but, although NGOs may be included in the private sector it is important to distinguish them from the profit seeking private sector, because they constitute a non-profit private sector (Meyer, 1992: 1115–26).

There has been a growing demand for tourist facilities that do minimal damage to the environment, and African wildlife holidays have proved particularly attractive in the ecotourism market. Campfire supporters hoped that the Communal Lands could take advantage of the burgeoning market in ecotourism. Revenue generation from wildlife utilisation takes consumptive and non-consumptive forms. Although non-consumptive tourism is favoured by donors and some NGOs, it is difficult to raise the kind of returns offered by sport hunting. This is because tourism requires some level of investment in the form of international standard accommodation, roads to carry viewing vehicles and other developments. Tourism-based Campfire programmes have developed, but they are only feasible where there is a significant scenic or sporting attraction, such as white-water rafting in the Zambezi River and Victoria Falls area. However, smaller-scale tourist facilities for Zimbabweans are run by communities. For example, Hurungwe District runs Sunungukai tourist camp, where local tourists and those from overseas can go on photo safaris. The community also provides basic accommodation in chalets and offers guided walks, fishing, visits to local villages and local handicrafts. Since there is no commercial operator involved, the community runs the chalets and retains all the revenue they generate.[28] However, tourism development in rural communities has been criticised. Critics have been concerned that the real profits have gone to international tour operators and hotel chains, rather than rural communities. Similarly, dependence on tourism means that rural people are at the mercy of an industry that is fickle, subject to changing fashions and easily disrupted by adverse publicity (Hall and O'Sullivan, 1996; Sonmez, 1998; Sonmez and Graefe 1998).

In contrast to photographic and cultural tourists, hunters require minimal development. This means that there is an immediate return from the trophy fee paid by the hunter to the community. A single foreign hunter can bring in up to seventeen times the foreign currency that an individual photo safari client brings in, and the impact on the environment is not as great. A single hunting party will create much less stress on the environment than numerous groups of tourists in vehicles interested in wildlife viewing. Hunting fees are the main source of revenue for Campfire areas. Elephants have attracted the largest trophy fees. Sally Bown, of the Zimbabwe Association of Tour and Safari Operators (ZATSO), stated that, in 1993, the fee for shooting an elephant in a safari area operated by the Parks Department was set at US$3000, whereas in a Campfire area anything up to US$9000 could be paid.[29] Pro-hunting lobbyists and Campfire advocates argue that hunting operators can provide quick and direct benefits to communities, with the minimum amount of investment and expenditure on their part. This, in turn, means that communities can see the link between wildlife and revenue immediately, whereas with tourist development the returns might be delayed (Jansen, undated). Hunting has been limited to small numbers of suitable species in order to ensure that wildlife is sustainably used. The Parks Department sets a hunting quota for each Campfire area, which is based on surveys of wildlife populations in those areas, but hunting quotas can be increased slightly by allowing PAC to be carried out by commercial safari operators.[30] Hunting rights are sold to safari operators, who pay for the right to hunt and the fees for each animal shot on the quota. For example, Tshabezi Safaris paid Muzarabani District Council the equivalent of US$108,025 in one year in hunting and trophy fees (Child, B., 1995d: 1). Yet rural communities do not choose the safari operator solely on the level of fees offered. In some areas, the community will choose a tender that offers high hunting and trophy fees but also offers incentives, such as building a road or lodges. In some Communal Areas, tenders were awarded on the basis of an existing good record of cooperation between an established safari operator and the community.[31] To support this kind

---

[28] Interview with Keith Madders, Director, Zimbabwe Trust (UK), 12.10.94, Epsom, Surrey; *Hurungwe District Campfire Newsletter* no. 5, Dec. 1993.

[29] Interview with Sally Bown, ZATSO, 9.3.95, Harare.

[30] Interview with Brian Child, head of the Campfire Coordination Unit, DNPWLM, 16.5.95, Harare.

[31] Interview with Sally Bown, ZATSO, 9.3.95, Harare.

of development, guidelines were drawn up by government agencies and NGOs involved in Campfire on how to run joint ventures between District Councils and private operators in order to ensure greater local control over hunting firms (Jansen, undated: 4).

The importance of safari operators in revenue generation has not been unproblematic. Some communities attempted to set up their own safari operations, so that all the fees and profits could be retained by the community, but, overall, these have not been successful, because the councils lacked the necessary expertise. For example, the community-run safari operation in Guruve failed to compete with the returns offered by commercial operators. Keith Madders suggested that the failings were due to lack of experience and the expectations or perceptions of potential clients. Part of the hunting experience is the night life – that is, swapping stories around the fire, where overseas clients appreciated someone well travelled (and white). Rural people could not compete with the services offered by commercial operators, such as catering for the dietary preferences of overseas clients, equipment, the social life and tented accommodation.[32] The fact that the safari industry is white-dominated has led to criticisms of Campfire. Critics suggested that Campfire has protected white safari operator interests over those of rural people, and that safari operators made very large profits from Campfire, which explains their vocal support for the policy (Murombedzi, 1992).

The role of hunting operators has revealed deep fissures in the conservation community and it has indicated that conservation is always about articulation of a particular political ideology, rather than a matter for a pure, non-political science of biological conservation. The question revolves around whether hunting can be considered as a conservation tool and, if hunting is a tool, whether it is ethical to use it. Campfire's reliance on hunting for revenue has brought the conflict between sustainable utilisation and preservation into sharp relief. Reliance on sport hunting means dependence on overseas markets, which have been threatened by various moves to disallow trophies from entering the USA, inspired by the animal welfare lobby. Most sport-hunting clients are Americans and, if they were not allowed to bring trophies home, they might not be willing to pay the high fees that Communal Areas now rely on.

Supporters of Campfire have criticised the international ivory ban as detrimental to community-based conservation. The international ban on the ivory trade in effect reduced the yield per hectare of wildlife utilisation schemes.[33] Almost half of the 35 tonnes of ivory in Zimbabwe's central store came from Communal Areas, representing an opportunity cost to communities of US$3.5–7.5 million if ivory is worth US$200–400

---

[32] Interview with Keith Madders, Director, Zimtrust (UK), 12.10.94, Epsom, Surrey.
[33] *Sunday Mail* 8.7.90 'Ban on Ivory to Affect Farmers'; *Herald* 16.12.94 'Allow Trade in Legally Obtained Ivory and Rhino Horns: Campfire'.

per kilo (prices based on offers made to the Parks Department from various sources in the post-ivory ban period) (Child, B., 1995a). Consequently, agencies involved in Campfire have devoted time and effort towards political lobbying, aimed at keeping existing trade links open (such as markets for sport-hunted ivory), as well as trying to reopen ivory trading.

The ivory issue and sport hunting raised a number of international issues. Through the ivory ban, the cultural attachment to non-consumptive use of wildlife in the industrialised world directly affected rural people in Zimbabwe. Marshall Murphree argues that the costs of maintaining African wilderness to urban middle-class and upper-middle-class people in the West is relatively low, and it is hard for those groups to understand the enormous cost paid by rural people to maintain that wilderness if hunting for trophies is outlawed by the international community.[34] In many ways, this difference is a matter of international perspectives clashing with local perspectives. Supporters of Campfire argue that Western preservationism is ultimately self defeating, since it negates the role of local people in wildlife conservation. In addition, Campfire is central to the Zimbabwean Parks Department's political lobbying in the international arena for a reopening of the ivory trade. It constitutes an important means of justifying sustainable utilisation to potentially competing domestic political interests and highly critical international interests. Campfire constitutes an area of policy where there is a clear focus on the political nature of conservation, in the face of rhetoric that claims conservation is about the science of saving animals and not about politics. Western animal rights and animal welfare lobbies have argued on moral and ethical grounds that Campfire relies on cruel practices, which cannot be supported, even if they contribute to poverty alleviation and rural development. It is important to note that Campfire has been successful in generating large amounts of revenue and has never been dependent on the ivory trade, despite claims by some Zimbabwean conservationists in 1989 that Campfire would fail if the ivory ban were implemented.

The involvement of varied local NGOs demonstrates Campfire's ability to weld together potentially competing interests. NGOs have been critical actors in the establishment and development of Campfire. Initially, the Parks Department envisaged a state-run Campfire Agency, which would recruit communities to the programme, implement the preliminary stages of Campfire and then reduce its involvement in the final transition stage, where communities would take full control for themselves (Martin, 1986: 30–2). However, it was soon clear that the Parks Department did not have the breadth of expertise or the levels of finance necessary to get Campfire started. It was hoped that donors might meet the full costs of a foundation grant for Campfire and recurrent expenditure in the first few

---

[34] Interview with Prof. Marshall Murphree, CASS, University of Zimbabwe, 18.7.95, Harare.

years. The gap in finance, research and expertise was filled by a number of NGOs, which came together to form the Collaborative Group (CASS/WWF/Zimtrust, 1989), which is now heavily involved in political lobbying on behalf of Campfire. The Collaborative Group consists of the World Wildlife Fund (WWF), Zimtrust, the Centre for Applied Social Science (CASS) at the University of Zimbabwe, the Campfire Association, the Parks Department and the Africa Resources Trust (ART). It demonstrates how Campfire has managed to weld together often diverging interests and perspectives. The Collaborative Group incorporates a number of professional and academic disciplinary interests, which span community development, wildlife managers, donors and aid workers.

Zimtrust is essentially a rural development agency and its aims are the relief of poverty and improvements in the quality of life in rural areas. Zimtrust's role in Campfire is to assist in institution building. It supports communities at ward and district level in their efforts to create the required mechanisms to be granted appropriate authority status by the central government (Zimbabwe Trust, DNPWLM and Campfire Association, 1990). CASS provides research support for the Collaborative Group, concentrating on the sociological, planning and project implementation aspects of Campfire (CASS/WWF/Zimtrust, 1989). WWF-Zimbabwe has been important in terms of capital investment and providing ecological surveys of potential and existing Campfire areas. The Multispecies Animal Production Systems (MAPS) project began as surveys and research for wildlife utilisation on commercial farmland and, with Campfire, it was extended to Communal Lands. MAPS staff have also been involved in a significant amount of extension work in rural areas. However, the project has steered away from an implementational role, which might be viewed unfavourably in political terms by other NGOs that are involved (Martin et al., 1992: 24–30). Finally, ART and the Campfire Association are involved in public relations. ART has tended to concentrate on international public relations and has promoted the concept of Campfire around the globe. It produces publicity material for Campfire and its agenda is to ensure that markets for sport hunting and hunting trophies remain open in Britain and the USA.[35] The Campfire Association was formed partly to lobby in favour of the interests of rural people's interests at the Convention on International Trade in Endangered Species (CITES) conferences. The Association attended the critical 1989 Conference and was the first interest group to represent indigenous people at this forum (Metcalfe, 1992c: 13). The Campfire Association is also a representative umbrella organisation for Campfire areas, intended as a channel to articulate and promote the interests and needs of rural people at a national level. The Campfire

---

[35] Interview with Jon Hutton, Director of Projects, ART, 13.3.95, Harare; interview with David Cumming, Director, WWF-Zimbabwe, 22.2.95, Harare.

Association and ART are definitely political and campaigning organisations, which lobby to gain domestic and international acceptance of Campfire, in particular, and sustainable utilisation, in general.

The Collaborative Group has opened up opportunities for NGOs to become deeply involved in rural development and conservation. It is significant that ecology-centred NGOs, such as WWF, traditionally work alongside hard-line rural development NGOs, such as Zimtrust. This is largely due to the fact that Campfire suits a number of different interest groups, including those interested in the environment, in wildlife, in black empowerment and processes of indigenisation, in rural development and in decentralisation of power.

State interests and involvement in Campfire

In 1982, an amendment was added to the 1975 Parks and Wildlife Act that allowed communities to be designated as appropriate authorities through their District Councils (Zimbabwe Trust, undated b; Lopes, 1996: 148–59). The District Council is, in principle, an accountable institution. Depending on one's perspective, District Councils can be viewed as a controlling arm of the state in rural communities or as truly representative institutions looking after local interests (Hill, 1996; Kepe, 1997; Neumann, 1997). The way in which the District Council is viewed varies according to district.

Devolution of powers to District Councils has not been unproblematic. Campfire has run into difficulties in ensuring that the District Council constitutes a representative institution for diverse community interests. It is difficult for a group of elected and non-elected officers to represent all the interests of a community. This problem is compounded by the differences in background and in the interests of the councillors in relation to the rural people they represent. Although the District Council is an elected body, the secretariat that runs it is not. The secretariat is meant to be part of the local government structure and thus serve the interests of the rural community, as expressed by the District Council elected officers. Yet one of the first problems that Zimtrust encountered was with the secretariat, because the people who made up the secretariat were setting the agenda for the council. Members of the secretariat had all lived in an urban setting and been to university. The agenda was, then, one that reflected the interests of a small urbanised élite in the locality and it closely followed the concerns of central government, rather than those of the rural community.[36] This compounded local people's view that the District Council was an administrative arm of the state and reduced the legitimacy of the council in the local community (Thomas, 1991: 18–20; Hill, 1996; Kepe, 1997; Neumann, 1997).

[36] Interview with Keith Madders, Director, Zimtrust (UK), 12.10.94, Epsom, Surrey; interview with Brian Child, head of the Campfire Coordination Unit, DNPWLM, 16.5.95, Harare.

Devolution of powers has not been matched by further devolution of revenues to the ward and village level in some areas. The failure of a number of District Councils to meet their obligations by devolving revenue can be explained as partly the result of underfunding from central government. A second reason is the attempts by the Ministry of Local Government, Rural and Urban Development (MLGRUD) and its political allies to obtain control over Campfire revenues for their own purposes. According to Brian Child, more revenue has reached the grass-roots level and, over the period 1990 to 1993, the overall amount of revenue retained by District Councils decreased from 38 per cent to 21 per cent (Child, B., 1995a: 20). However, some councils have been accused of retaining too much wildlife revenue, and part of this problem is that the definition of how much money the District Council can retain is blurred. Policy drafts by the Parks Department stipulate that the District Council cannot keep more than 15 per cent of the revenue to cover their overheads, and that 35 per cent can be spent on wildlife management activities, such as law enforcement and monitoring. The difficulty has arisen in deciding which groups should spend on wildlife management activities. District Councils retained the revenue for wildlife management, rather than giving it to producer communities. The District Council can, in theory (and, in some cases, in practice), retain 50 per cent of wildlife revenues (15 per cent for overheads and 35 per cent for wildlife management activities) (DNPWLM, 1991: 1–4). The goal for Campfire is that producer communities should receive 80 per cent of all the revenue generated, with the District Council retaining no more than 20 per cent (Child, B., 1995b: 14).

The majority of District Councils have been unwilling or unable to decentralise powers to smaller institutions at the ward and village level. The difficulty is that an institution capable of being granted appropriate authority does not yet exist at the ward and village level. Brian Child argued that it was not possible to decentralise powers to something that simply did not exist.[37] This lack of further decentralisation has become informally known as Murphree's Law, which notes that each level of an organisation attempts to wrest control from the levels above it and resists the devolution of power and control of funds to the levels below it (Murphree, 1995). Indeed, the 1994 Rukuni Commission stated that senior authorities in MLGRUD down to the smallest Village Development Committees (VIDCOs) seemed to believe they had *de jure* exclusive authority over Communal Lands (Government of Zimbabwe, 1994b: 28). There has certainly been a build up of vested interest groups at the District Council level, which stifles the powers of interest groups at the ward and village level (Hasler, 1990: 1–6). Murombedzi suggests that, far from decentralisation, Campfire has resulted in recentralisation of power

---

[37] Interview with Brian Child, head of the Campfire Coordination Unit, DNPWLM, 16.5.95, Harare.

and resources at the District Council level. The central government has merely been replaced by an arm of the state in the form of local government (Murombedzi, 1992; Hill 1996; Kepe, 1997; Neumann, 1997). The conflict surrounding disbursement of revenues demonstrates that the District Council has obtained powers over a certain area and has been unwilling to relinquish that control.

This debate is part of the wider political struggle for resources from wildlife. The Campfire Coordination Unit estimated that, by 2005, Campfire could earn Z$100 million per annum for Zimbabwe (DNPWLM, 1994: 4). The potential of Campfire to be a major revenue generator has attracted the attention of a number of political groupings, each with their own interest in how the money from wildlife schemes should be spent. Murphree argues that Campfire has overconcentrated on pragmatic strategies and so masked deeper political and economic implications in development policy, such as decentralisation and participation. As a result, Campfire implementers have failed to understand the political context of Campfire and the political threats to it (Murphree, 1995). Those who oppose Campfire have important and prominent political allies, whose interests lie in wresting control over wildlife from rural people.

Recentralisation of powers and revenue at the District Council level is partly the result of the national process of economic adjustment. The Economic Structural Adjustment Programme (ESAP) resulted in budget cuts for all government departments. The District Councils were aware of the need to supplement their diminishing budget allocations from the central government, and so the District Council and its officers have been interested in obtaining a proportion of wildlife revenues.[38] This lack of finance was a major factor in District Councils failing to disburse Campfire revenues. There were accusations that some District Councils retained the revenue and used it for other areas of council expenditure, such as fax machines and vehicles for the councillors.[39] This problem was aggravated by the amalgamation of Rural Councils and District Councils into Rural District Councils in 1993. The new councils were the result of a merger between Rural Councils, which were largely representatives of white commercial farming areas, and District Councils, which represented black Communal Lands. The details of how the new Rural District Councils would be financed were not fully developed at the planning and amalgamation stages, and this raised questions about their financial viability in the long term, especially since they merged wealthy commercial areas with much poorer Communal Lands (Roe, 1992: 1–5). The lack of finance for day-to-day council operations meant that wildlife

---

[38] Interview with Keith Madders, Director, Zimtrust (UK), 12.10.94, Epsom, Surrey.
[39] Anonymous interviewee; see also *Herald* 5.5.95 'Villagers Express Fears Over Campfire Benefits'; *Daily Gazette* 13.10.93 'Minister Urged to Investigate Councils'; *Daily Gazette* 2.10.93 'Call to Pay Wildlife Proceeds'.

revenues were perceived as a means of supplementing council budgets. A number of NGOs involved in Campfire viewed this retention of funds by District Councils as a very large tax on wildlife activities.[40] In the long term, such a high level of tax carries the threat of making wildlife less profitable, compared with agricultural produce and livestock, which are not subject to taxation.

The relationship between Campfire areas and central government is a complex one and it is often very strained. The generation of large amounts of revenue that are never passed through the central treasury has caught the attention of certain officials and departments. Currently, the revenue in Campfire falls just below the threshold for taxation by the central government. Brian Child estimates that, in 1994, Campfire generated approximately Z$20 million. The revenue generated by Campfire schemes has grown dramatically since 1989. In future years, Campfire could fall into the tax bracket.[41] Marshall Murphree noted that Campfire was entering a 'catch 22' situation. In the initial stages of Campfire, rural people had to be convinced of the economic value of wildlife and revenues were relatively small. The catch comes when wildlife becomes so valuable that state or substate organisations and the private sector become interested in wildlife utilisation. The danger is that these interests could seek to pull back the rights of rural people over wildlife for themselves.[42] This increase in revenue from wildlife utilisation has attracted the attention of elements in the central government. Two of the most outspoken critics of individual payments have been the late Vice President, Joshua Nkomo, and the Minister of Local Government, Rural and Urban Development, Joseph Msika.[43] Local government would benefit from communities spending their revenues on district development projects, such as roads and clinics. This is because, as with all government departments, MLGRUD has been subject to budget cuts. If Campfire revenues were spent on projects that would normally be the responsibility of local government, the authorities would be released from a large financial burden and still be able to present themselves as working for the benefit of rural people. Joshua Nkomo's concern over individual payments also partly arose from the continuing growth of Campfire revenues, which avoided the central treasury. Nkomo's opposition to individual cash payments to villagers and the devolution of decision-making powers to the local level is partly inspired by a need to resist further decentralisation of wildlife management. A number of those involved in Campfire and conservation in general

[40] Interview with David Cumming, Director WWF-Zimbabwe, 22.2.95, Harare; interview with Keith Madders, Director, Zimtrust (UK), 12.10.94, Epsom, Surrey.
[41] Interview with Keith Madders, Director, Zimtrust (UK), 12.10.94, Epsom, Surrey.
[42] Interview with Prof. Marshall Murphree, CASS, University of Zimbabwe, 18.7.95, Harare.
[43] *Daily Gazette* 28.9.93 'Minister Threatens Intervention'; *Daily Gazette* 13.10.93 'Minister Urged to Investigate Councils'.

commented that interference from central state agencies and certain central government figures was detrimental to the success of Campfire.

Finally, the fundamental aims of Campfire are at odds with the stated policy of the central government. Campfire espouses the rights of local communities and decentralised control. In contrast, the central government claims to be Marxist and in favour of centralised control of all decision-making. The strained relationship between Campfire and central government conforms in part to Murphree's Law, since District Councils resist control from the top and the central government attempts to slow further devolution of powers to the local level. In addition, the amount of revenue generated by wildlife means that in some areas Campfire institutions and the Campfire Association carry more legitimacy for local people than central government institutions. Consequently, Campfire-related agencies have been viewed as a threat by existing political authorities. Non-elected government bureaucrats, who openly suggest and promote new policies, such as Campfire, have been perceived to be acting as politicians and therefore as critics of ZANU-PF for not suggesting the policy in the first place (Roe, 1992). In addition, rural MPs have been elected by promoting the Campfire message. For example, Border Gezi (the MP for Centenary) is from the Communal Lands himself and was the first politician to be elected on a Campfire platform.[44] The election of a number of MPs on Campfire platforms has assisted the ability of Campfire agencies to articulate explicit political support for utilisation. These Campfire MPs have made extremely critical comments about the government's attempts to force rural people to spend wildlife revenues solely on development projects. For example, the MP for Guruve, Ephraim Chafesuka, warned the Minister of Local Government, Rural And Urban Development, Joseph Msika, not to interfere in genuine devolved decision-making by rural people.[45] The election of such pro-Campfire MPs from rural areas has strengthened the position of Campfire and raised its profile in national politics.

Campfire constitutes the central justificatory focus for sustainable utilisation in terms of domestic environmental politics and international environmental politics. On the domestic and international stage, critics have found Campfire the most difficult aspect of sustainable utilisation to argue against. Supporters of Campfire have used the argument that wildlife revenues are distributed to local people, who decide how to spend them to enormous effect, and they can present detractors as misanthropes, who seek to deny poor local people their proper right to use their own resources to improve their quality of life. As such, Campfire forms

[44] Pers. comm. Peggy Allcott, WWF Africa Section, Godalming, Surrey; *Herald* 9.9.92 'MP Hits at Council'.
[45] *Daily Gazette* 28.9.93 '*Minister Threatens Intervention*'; *Financial Gazette* 1.12.94 'Warning on Campfire Cash'; *Herald* 9.12.94 'Chafesuka Slams Campfire Plans'.

the centre-piece of Zimbabwe's argument for sustainable utilisation within the domestic political context and at an international level.

At a national level, there are a number of groups that benefit from Campfire, but equally there are interest groups that view Campfire as a threat. Campfire has managed to negotiate the potential differences between its supporters. Rural-development NGOs, hunting interests, private tour operators, wildlife conservationists, black empowerment groups and academic researchers have all cooperated in Campfire, because each group derives some benefit from it. Yet Campfire faces significant political threats. It has been such a success in generating revenue for rural people that it has caught the attention of cash-strapped ministries and of a centralised government concerned with its decentralisation and empowerment processes. First, the comments of the Minister for Local Government, Joseph Msika, reveal that the MLGRUD perceives Campfire as a chance to obtain control over much needed revenues. The disbursement of wildlife revenues to individual households in rural areas has been viewed as a waste of money, which could be profitably spent by the ministry on local administration and development projects. However, such a development would negate the fundamental principle of Campfire – that rural people must be allowed to decide how wildlife revenues are spent. Nevertheless, since Campfire is partially dependent on an arm of MLGRUD, the District Council, the threats from the parent ministry constitute a serious problem for the policy. The support from Nkomo for the Ministry's position demonstrates the level of concern in central government about a programme that involves such radical rural reform and decentralisation of powers to areas that the ruling party was once able to control.

The international political context has proved equally significant for the future of Campfire. It is a policy that brings the differences between environmental ideas and disputes between factions of the conservation community into sharpest relief. The international community has been more influenced by preservationist theories, to the extent that international environmental NGOs have become concerned with animal-welfare issues, which, in the long term, can conflict with conservation objectives. NGOs have campaigned against Campfire on the grounds that it relies on cruel practices, such as the cropping of wildlife for meat and sport hunting. However, the continuing development of ecotourism and photographic safaris in Campfire areas has reduced this reliance on hunting. Anti-Campfire groups have also pointed to the arguments for reopening the ivory trade as a reason to oppose the programme. The difficulty for Campfire is the importance of donor funds not just for wildlife but for other programmes. International donors based in Western states, which are influenced by animal welfare, could be pressured to reduce aid to countries that allow sport hunting, ivory trading and utilisation of wildlife.

6

Buying Influence? | The Politics of Donor & NGO Involvement

The use of non-governmental organisations (NGOs) by governments and donors to channel aid increased dramatically over the course of the 1980s and 1990s. The donor community's reliance on NGOs to deliver aid and development projects, in turn, allowed NGOs to extend their influence in the developing world. The diplomatic space left by the end of the Cold War provided an opportunity for global NGOs to direct the international agenda with regard to particular political issues, such as human rights and the environment. Equally, industrialised states appeared to regard NGOs as more effective channels for development aid and conservation projects than governments in developing states (Meyer, 1992; Clapham, 1996; de Waal, 1997). Global NGOs were relatively new and clearly promoted environmental care at a time when other organisations and states seemed unaware of the level of public concern about the environment. Since then, environmental NGOs have proliferated, and concern with the environment has reached mainstream political parties, academics, business and government policy. The 1980s witnessed an increase in the bargaining power of these organisations. Their ability to raise funds for projects all over the world was due to changes in the 1980s, which resulted in increased professionalisation and commercialisation of fund-raising activities. Consequently, NGOs controlled large blocks of funds, which could be released for the right project. International NGOs such as Greenpeace and the World Wide Fund for Nature/World Wildlife Fund (WWF) have a global public profile and they control large amounts of revenue donated by members and supporters, mostly in Western states. The increased ability of NGOs to influence the actions of states and multinational corporations was demonstrated by the campaigns against Shell over Ogoni rights and the oil industry in Nigeria.[1] Consequently, NGOs have become a new political force, able to

---

[1] *Economist* 20.7.96 'The Fun of Being a Multinational' pp. 63–4; *Independent* 17.11.94 'What On Earth Do They Do Now?'; *Independent* 30.10.96 'A Professional Tug at the Heart'; de Waal (1997).

influence governments and business through campaigning and funding conservation programmes. This chapter will explore the role of NGOs and donors in conservation policy in Zimbabwe. First, it will examine the role of local NGOs in conservation policy and the legislation that governs and controls them. Secondly, it will consider the role of international NGOs and will explore the differences between international and local NGOs. It will assess the problems international NGOs have in communicating their policies to their memberships, which is highlighted by the case of WWF, and will examine the influence of animal rights-based organisations on specific policy areas, such as elephant culling and translocation. Finally, it will investigate the importance of other donors, such as government-to-government aid and World Bank funding for specific projects.

The Zimbabwean NGO environment

Bratton suggests that NGOs can influence public policy choices under certain circumstances. For example, they must represent a homogeneous clientele with a unified agenda, their relationship with central government must be one of negotiation rather than confrontation and they must have access to the necessary resources to carry out their aims (Bratton, 1990). NGOs have become critical actors in conservation policy. They are often at the interface between differing interest groups, such as between government and community, between donor and beneficiary and between local, regional, national and international perspectives (Metcalfe, 1992a: 14). Donors, whether NGOs, governments or international financial institutions, have an ability to exert pressure on states to continue with current policy or to change policy direction. Meyer argues that NGOs provide services for international donors, such as governments and the World Bank, as aid conduits. As such, their programmes reflect donor concerns, rather than local ones (Meyer, 1992: 1120–3). Since donors can offer desperately needed conservation aid, in many cases states are obliged to adhere to the particular conservation philosophies preferred by donors. Lal argues that global environmental organisations have instituted a new form of green imperialism, which inherited the characteristics of previous colonial movements from Western states (Lal, 1995). Furthermore, Vivian suggests that NGOs suffer from a magic-bullet syndrome, in that they are perceived to be able to carry out multiple tasks and satisfy multiple targets, and they must demonstrate their success in order to secure further donor funding for their projects. Consequently, NGOs can be secretive about the actual results of their activities (Vivian, 1994: 187–90; Hanlon, 1996; de Waal, 1997). However, it is clear that NGOs often operate in a perceived social, economic and political vacuum and, as a result, their programmes and projects may fail to meet their objectives, once local factors have begun to exert an influence.

Numerous national and international environmental NGOs are involved in environmental projects in Zimbabwe. Their involvement ranges from support for local schemes to campaigning to change government policy. The tensions between a number of organisations and the Zimbabwean state over certain aspects of conservation policy reveal important differences in approach. It has caused pro-utilisation conservationists in Zimbabwe to accuse donors of a new environmental imperialism, claiming that those who offer conservation aid formulate rigid ideas of policies worthy of funding, regardless of the conditions that prevail within the state. As a result, conservation aid is often wasted on projects that are not politically, economically or socially sustainable, because they have been divorced from those conditions that have a crucial impact on how they perform.

The type of influence that donors have over public policy choices has led to confrontational relationships between local NGOs and the Zimbabwean government on the one hand, and international NGOs, on the other. Organisations that do not offer funding for field projects can still have an effect on policy through campaigning activities. Clearly, Zimbabwe's policy position is relatively unpopular with the constituencies of Northern donors that prefer preservation to sustainable utilisation. The relationship between Zimbabwe and external donors and conservation NGOs is a complex and interesting one. While some donors denounce Zimbabwe's policy, others quietly support it and provide funding for it.

NGOs are not a single homogeneous group, since they are divided by their conservation philosophy, their role as local or international organisations, their campaign focus and their relationship with the national governments under which they operate. Local Zimbabwean NGOs are almost exclusively in favour of sustainable utilisation of wildlife. The exceptions are rare animal welfare organisations, such as the Elephant Rose Foundation and the Southern African Group for the Environment (SAGE). Local NGOs are also defined by their relationship with the Zimbabwean government, which has restricted their autonomy with recent legislation. In contrast, international NGOs can often be far more autonomous in their actions. For example, campaigning rather than field organisations are able to denounce what they view as conservation failures, whereas those that have operations within a country have to consider the effect of denunciations on their projects. In addition, international NGOs are defined by their general attachment to animal welfareorientated ideas. There are exceptions, such as Oxfam and, increasingly, WWF, which are concerned with community-based conservation.

Wildlife conservation is a policy area that has historically been dominated by the white settler community in Zimbabwe and it provides an important political context in which environmental NGOs operate. The history of conservation has fed into the composition of local conservation

NGOs in the post-independence era. The majority of conservation NGOs are run by white Zimbabweans and their membership is mainly composed of white Zimbabweans. In general, wildlife conservation is perceived as an issue that is captured by the settler community. The racial overtones in conservation are apparent, and many of those interviewed clearly stated that they were hoping to increase black membership and black representation on executive boards, but admitted that they had experienced great difficulty in achieving this.[2] In recent years, there have been initiatives to increase participation by black Zimbabweans in wildlife conservation organisations. For example, the Communal Areas Management Programme for Indigenous Resources (Campfire) Association actively recruits black staff and attempts to represent an overwhelmingly indigenous constituency. This is because conservation as rural development strategy is perceived as an issue for the Communal Lands. Local NGOs have a distinct role to play in Zimbabwean conservation, which is mostly a supportive role, and in general they present a unified front on key conservation issues, notably Campfire, the ivory trade and sustainable use. However, their relationship with the central government is an ambiguous one, since, on the one hand, they support sustainable use, which also represents their own constituencies, but on the other hand they have been increasingly constrained by government legislation.

Zimbabwe has a strong tradition of local NGO involvement in environmental issues, including contributing to public discussion and fund-raising. Some are financed by donor funding or by membership fees and fund-raising events. Local NGOs also vary in their activities, which range from campaigning organisations to those that provide specific expertise. In general, they are all supportive of government policy towards wildlife. This tradition is evidently derived from the historic involvement of the white community in NGOs generally and the environment in particular, creating difficulties for political action in post-independence Zimbabwe, which will be discussed later. It does mean, however, that the role of NGOs in Zimbabwe is not restricted to liaising with international donors, but involves a complex relationship between external actors, local NGOs and the Zimbabwean state.

Zimbabwean NGOs that are involved in conservation vary considerably. They are involved in campaigning, specific projects and national policy programmes, such as Campfire, and also provide assistance for government agencies, but they have one common factor, in the sense that, in general, local NGOs are supportive of Zimbabwe's conservation policy. For example, NGOs whose constituencies and focus of interest differ widely are in favour of sustainable use. This support for

---

[2] For example, interview with Charl Grobbelaar, Chief Executive of the Zimbabwe Hunters Association, 12.2.95, Harare; interview with Randall Foster of the Wildlife Society, 22.5.95, Harare.

government policy has arisen out of the unique conservation history of Zimbabwe. The state and conservation organisations have had a lengthy commitment to sustainable use, because of the relative abundance of wildlife, the importance of making each sector of the economy pay its way and the (white) Rhodesian preference for outdoor leisure pursuits.

The nature of NGO support for government policy philosophy has been translated into material assistance for the Parks Department. In Zimbabwe, NGOs often perceive themselves to be filling a gap left by a government that has numerous pressures on its budget. Local NGOs play a key role in providing material assistance for government agencies by giving specific support and equipment to the Parks Department. For example, the Zambezi Society began an Adopt a Scout scheme to provide funding and equipment for Parks scouts in the Mana Pools National Park.[3] Similarly, the Voluntary Service is comprised of members of various Zimbabwean NGOs, who assist the Parks Department with routine duties and maintenance of equipment. In 1994, the first commercial tour operator, Frontiers Rafting, provided funding for the Mashonaland Branch of the Wildlife Society, acting as part of the Voluntary Service.[4] Local NGOs clearly recognise the problems associated with the Department, but prefer to provide services that would normally be the responsibility of central government, and this sharply contrasts with Western or international NGOs, which are more inclined to criticise central government and lobby for increased resources. Unlike NGOs in industrialised states, they tend not to criticise government policy philosophy or policy action but instead constitute an important support base for wildlife agencies, in terms of providing material support and campaigning on behalf of government policy in the international arena.

Bratton suggests that the ability of NGOs to influence policy decisions rests in part on their capacity for forming close personal ties with 'the powers that be' (Bratton, 1990: 111). This ability has proved especially important for Zimbabwean NGOs. A few local NGOs are highly critical of policy and of the government, such as SAGE and the Elephant Rose Foundation.[5] However, such exceptions are rare. In general, local NGOs lobby in a non-confrontational manner. Rather than staging demonstrations and publicity stunts to capture public and government attention, they engage in quiet diplomacy. Graham Child of SAVE described this as the shirt and tie approach to lobbying, commenting that he perceived this to be more effective than public confrontation and denunciation of government policy.[6] The effectiveness of this style of lobbying is partly due to the close personal and professional ties between the directors of

---

[3] Press release by the Zambezi Society 19.6.95 'Zambezi Society Sponsors Anti Poaching Scouts'.

[4] *Financial Gazette* 29.9.94 'Voluntary Service to Work with National Parks Department'.

[5] *Sunday Mail* 1.5.94 'Wildlife Relocation Backed'; *Sunday Mail* 24.4.94 'Management of Rural Wildlife Sparks Wrangle'.

[6] Interview with Graham Child, Director, SAVE, 27.2.95, Harare.

NGOs and the Parks Department. Graham Child is a former director of the Parks Department, Dr David Cumming of WWF-Zimbabwe was deputy director of the Parks Department, and Dick Pitman of the Zambezi Society was also a Parks employee. All three moved out of Parks at around the same time, during the brain drain of the mid to late 1980s, and went into the NGO sector.[7] Similarly, white Zimbabweans have retained a position of relative privilege in society, dominating conservation circles and having social and family links with the senior management of the Parks Department.

However, NGOs have managed to offer criticism of certain policy decisions, while remaining broadly supportive. This is highlighted by the Zambezi Society. Dick Pitman stated that the Society lobbied policymakers, which may or may not include the Parks Department, over what the Society considered inappropriate development for the Zambezi valley. The Parks Department has been generally in favour of the Society's lobbying, which includes criticising the development plans of other, potentially rival, Ministries.[8] Similarly, NGOs found that commercial business tended to be responsive to such lobbying, because it is potentially damaging for their public image. However, there is very little dissent over actual conservation policy. Rather, there has been some criticism of the government and the Ministry of Environment and Tourism for failing to allocate adequate funding to the Parks Department.[9] Local NGOs have found means of opposing the central government, but the increasing influence and financial power of NGOs has spurred new legislation aimed at restricting their autonomy.

In 1995, the government introduced far-reaching measures to further control NGOs. The Social Welfare Organisations Amendment Act was promoted as a means of preventing mismanagement and misappropriation of funding by NGOs. The Act was not aimed at conservation NGOs in particular, but its provisions did affect their day-to-day operations and their freedom to lobby. The provisions of the Act and the way it was introduced raised serious concerns about the authoritarian nature of the Act. First, NGOs were not consulted over the introduction of the Act, but were only informed of its existence from the news media, once it had passed its second Parliamentary reading (Zimrights, 1995a). Concerns were raised over the potential effect on NGOs of redefining them as private voluntary organisations (PVOs), which made them subject to taxation that could leave some NGOs bankrupt. However, the main fear expressed by NGOs was that the Act gave extensive powers to the Minister of Public

---

[7] *Financial Gazette* 20.11.87 'Seven More Senior Officers Leaving National Parks Department'; *Financial Gazette* 4.12.87 'Parks Deputy Director Takes Early Retirement'; *Financial Gazette* 3.6.88 'Crisis Point at National Parks as Twelve Senior Staff Resign'.

[8] Interview with Dick Pitman, Zambezi Society, 30.5.95, Harare.

[9] For example, interview with Julie Edwards, Environment 2000, 12.7.95, Harare; interview with Dick Pitman, Zambezi Society, 30.5.95, Harare.

Service, Labour and Social Welfare and that the Welfare Advisory Board was elevated to the status of a decision-making body under the direction of the Minister (Zimrights, 1995b). Under the Act, the Minister can suspend the executive of any NGO or certain members of it and replace them with the Minister's own appointees if it is believed that the NGO is failing to act according to its objectives. Similarly, the Minister does not have to state any reasons for the action and there is no provision for NGOs to appeal against decisions.[10] Local NGOs were concerned that such extensive powers could be used to effectively silence NGOs that are critical of government or policy. In addition, external donors had preferred to channel funding through NGOs, rather than the central treasury in order to ensure that the funds reached the target group. The Act ensured that the central government stretched its tentacles into those NGOs, which had the potential to make them less attractive to external donors who preferred what they saw as NGO independence from government interference (Zimrights, 1995b).

The government presented its own set of reasons for introducing the Act, while NGOs that opposed it viewed it quite differently. The government claimed it was a measure that would empower NGOs and deal with the problems of mismanagement. NGOs claimed that the growth of civil society represented by the increasing number of NGOs and their involvement in advocacy work was of concern to the government, which introduced measures to control them.[11] What is clear is that the extensive powers granted to the Minister were in contravention of the principles of transparency and accountability that had been recently espoused by the Zimbabwean government. Indeed, it was suggested that the Act was rushed through Parliament in the last session before the general election to prevent dissent among MPs, and NGOs that opposed the Act believed that the MPs were concerned only that they would be deselected by ZANU-PF as election candidates if they were seen to oppose any government measure in the run up to the election in April 1995.[12] There were fears that the new Act would eventually reduce the amount of criticism publicly levelled at the government, since NGOs would be concerned that their critical comments would result in NGOs being dissolved or taken over by the Welfare Advisory Board. The 1995 Act further restricted the autonomy of local NGOs and, in many ways, it increased the historically supportive role that conservation NGOs have had in Zimbabwe, but it expanded that role by applying negative pressure to NGOs.

[10] Welfare Organisations Amendment Act no. 6 1995; see also *Horizon*, July 1995, 'State Seizes Control Over Local NGOs'.

[11] Interview with Ozias Tungwarara, Director of Zimrights, 12.7.95, Harare.

[12] Interview with Daniel Mtetwa, Director, Zimbabwe National Environmental Trust (Zimnet), 15.3.95, Harare.

International NGOs

International donors have had an influential role to play in wildlife policies across sub-Saharan Africa. King suggests that the new scramble for Africa in the 1990s is the scramble of aid agencies and donors for projects that will bring them credit, fulfil their legislative mandate and enable them to dispense funds without incurring criticism at home (King, 1994: 32). The operation of global donors is closely related to the conservation ideology they are attracted by, and international donors and NGOs tend to adhere to a rather different conservation philosophy from that of southern African states. Western attachment to ideas of animal welfare has had an effect on the direction of international NGOs. In addition, their access to funds and the nature of their support base tends to make them more autonomous than local NGOs. As a result, they are able to fulfil their own agenda, rather than taking on an existing government one. However, the level of autonomy varies according to the type of organisation, because some are campaign-based organisations, while others concentrate on field projects, and some wield power with large amounts of aid money, while others are small organisations with little finance to offer. International NGOs tend to be more critical and more concerned that their donor funds be spent on specific types of conservation programme. This occasionally results in difficulties and resentment from the state and local NGOs that the international organisation is attempting to work with.

International NGOs interact with the conservation community in Zimbabwe in a number of ways. They can provide support for local NGOs or join them in a partnership, they can assist the national government with funding or equipment or they can campaign against the government in international forums. Rowan Martin stated that he felt that the environment had become big political business and that this development brought differences in conservation opinion to the fore. Consequently, preservationists and animal welfarists have captured the balance of political power, while presenting advocates of sustainable utilisation as extremists (Martin, 1993). It is in the campaigning arena that the tension between international NGOs and the Zimbabwean government has been at its height.

Conservation NGOs have campaigned by appealing to ethical and moral arguments to carve out a moral space in international forums, such as the Convention on International Trade in Endangered Species (CITES) (Nadelman, 1990: 481–3). However, commitment to a particular ideology has implications for NGO relations with developing states. One of the problems that international NGOs face is the historical and cultural differences between the Western industrial societies that provide funding and support and the societies in which they have programmes. For Zimbabwe, a major difference is the relationship between land ownership and wildlife conservation. In the USA, for example, ownership of wildlife has historically been vested in the state, while, in Europe, rural landowners had some control over wildlife, but in the twentieth century control was transferred to the state. Martin (1994b: 4) suggests that the transfer of ownership rights to the state was the result of political lobbying by the new urban majority (see also MacKenzie, 1988). In Zimbabwe, ownership of land was split three ways by the colonial government, between the state, white commercial farmers and small scale (black) communal farmers. The ownership of wildlife was vested in landowners by the 1975 Parks and Wildlife Act, which included commercial farmers, and by the 1980s was extended to the District Council as the appropriate authority for communal farmers. Pro-utilisation conservationists argue that Western conservation organisations have failed to recognise the importance of this difference, since they argue that landowners expect whatever they produce on their land to be economically viable and preservation tends not to be. As a result, Zimbabwean conservationists have found themselves increasingly arguing against Western scientists, conservationists and governments for the right to use their own wildlife (Martin, 1994b: 9).

Such differences have led to complaints about international NGOs working in Zimbabwe being remote and unaccountable to local people or local NGOs, and that they denounce sustainable utilisation on moral grounds without fully understanding it (Metcalfe, 1992a: 59). International NGOs often treat wildlife as though it existed in a vacuum, divorced from the social, political and economic environment in which it lives. Wildlife is not tied to issues of the land, including who owns it, who uses it and who benefits from the resources on it.[13] Western conservationists tend to promote deep-green-orientated preservation of wildlife for its aesthetic and ethical values. The attitudes of donors and conservationists often closely mirror those of the industrialised world. This has led to the imposition of wildlife preservation plans that ignore the genuine needs of rural people in particular (Zimbabwe Trust, DNPWLM and the Campfire Association, 1990: 4). This preservationist attitude is often explained by Zimbabwean conservationists as the outcome of societies that have destroyed their large and dangerous wildlife and are relatively divorced from the natural world.[14]

These divergent concepts of the environment have extended into donor and NGO policy, to the extent that these fundamental differences in perceptions of the environment lie at the heart of disputes between Zimbabwe and some of its donors over conservation. In developing countries, conservation often means large investments of land and finance that are demanded by other development programmes. The influence of NGOs is felt to be almost overpowering, especially in the case of elephant

---

[13] Interview with Rowan Martin, Assistant Director: Research, Department of National Parks and Wildlife Management (DNPWLM), 29.5.95, Harare.
[14] Interview with Willie Nduku, Director, DNPWLM, 7.7.95, Harare.

policy. Thus, Mick Townsend of the Wildlife Society of Zimbabwe stated that external funds were always most welcome, but that the strings that were frequently attached to such funds were not. He viewed the role of NGOs as restrictive and stated that Zimbabwe's elephant policy could be jeopardised by the particular objectives of a single NGO.[15] A number of Zimbabwean NGOs argue that Western conservation organisations should consider whether they are justified in imposing their views on people who live with wildlife under very different circumstances (Zimbabwe Trust, undated a: 9). Real fears do exist in Zimbabwe that the country will become like other African states, where it is perceived that conservation policy is out of the hands of government and in the hands of NGOs. To combat this, government policy is to direct international NGO support through local NGOs or local chapters of international NGOs, in the hope that this would ensure Zimbabwean control over the activities of international NGOs (Martin et al., 1992).

However, the charges of cultural or ecological imperialism levelled at international NGOs serve other political purposes. It allows the Zimbabwean Government to present itself as not bowing to the interests of outsiders and as being in control of policy decisions. Accusations of eco-imperialism are often used in the hope that this will make Western liberal NGOs more cautious and more sympathetic. International NGOs have a difficult relationship with the states they campaign against and also those they attempt to work with. In the case of Zimbabwe, its policy philosophy is not popular with the majority of green organisations and this means donors can choose to support certain policies and withdraw funding from others as a method of punishment or a demonstration of displeasure.

The process of globalisation in the post-Cold War era has ensured that NGOs play a crucial role in African conservation practice. The provision of funding and support from international NGOs follows the increasing pattern of allowing non-state actors to conduct relations with African states and government agencies (Clapham, 1996: 163–78). NGOs and donors alike view African states as incapable of ensuring that aid reaches the target areas or groups. As a result, a number of NGOs refuse to provide funding for the Parks Department in Zimbabwe, since the perception amongst NGOs is that giving money to the Parks Department will ensure that it does not reach conservation initiatives in the field. This is because all cash donations to the Department must be handed to the central treasury first, where they are often lost to other ministries. This view was reinforced by the political power struggle within the Parks Department and between it and other ministries. It became clear to potential donors that aid for wildlife conservation could not be divorced from the wider political situation in Zimbabwe.

However, most international NGOs and donors have found that selective collaboration with national governments has proved more

[15] Editorial, *Zimbabwe Wildlife* Jan-Mar. 1994, p. 5.

productive than confrontation. In this way, they can influence the policy environment even if they are unable fully to control it (Bratton, 1990: 96). NGOs differ in the type of assistance they give to the Parks Department. The size of the organisation can be an important determining factor in relationships between states and NGOs and between various NGOs. For example, the relationship between large and smaller conservation organisations is often strained. John Gripper of the Sebakwe Black Rhino Trust, suggested that small conservation charities face problems because the public perception is that large ones, such as WWF, will do everything that is necessary. In addition, when a small charity is established, it can be viewed as a threat by larger ones, since in theory there is only a certain amount of money that the public will give to conservation, and each NGO takes part of it.[16] However, small NGOs can also present themselves as pure and dedicated to giving all donations direct to conservation. For example, Tusk claims that 92 per cent of all funds raised in 1992–3 were made directly available to conservation work in Africa and that Tusk has voluntary representatives in Africa, thus ensuring projects are efficient and cost-effective.[17] In contrast, bigger organisations clearly have a large salaried staff and higher maintenance costs. In the mid-1990s, global NGOs, such as WWF, Greenpeace and Friends of the Earth, were forced to restructure and make staff redundant because of falling incomes. With the Northern states in recession at the turn of the decade, green organisations began to wonder if the environmental bubble had burst. The proliferation of the numbers of environmental NGOs also meant that large NGOs were worried that they no longer commanded the centre stage on green issues.[18] As with local NGOs, a number of international organisations often provide support for the Zimbabwean Government, and assistance is provided in the knowledge that the Department is underfunded and short of equipment to carry out basic tasks. For example, the British-based Conservation Direct pools the skills of volunteers into a 'green army' to carry out specific conservation projects in Zimbabwe.[19] Support for anti-poaching patrols and Parks rangers is especially offered by international NGOs (Cunliffe, 1994). This allows the NGOs to present themselves as doing something to halt the extinction of endangered species and it is a high-profile project, which attracts donations from members in Western states.

There are various reasons why NGOs provide finance for the projects that they do fund. Large conservation organisations tend to fund high-profile and single-issue projects or campaigns. High-profile projects generate publicity for the organisation as well as the project; they capture

---

[16] Interview with John Gripper, Sebakwe Black Rhino Trust, 14.10.94, Ascot-Under-Wychwood.

[17] *Tusk Talk* Summer 1994.

[18] *Independent* 17.11.94 'What On Earth Do They Do Now?'

[19] *Herald* 22.7.93 'Wildlife Body Set Up in the UK'.

the public imagination and so can bring in revenue through voluntary donations for the project. For example, NGOs will fund moving a captive dolphin to a rehabilitation centre or, as in Zimbabwe, the relocation of black rhino overseas. Care for the Wild and USAID, for example, were actively involved in funding the highly publicised elephant translocation exercise in the 1990s.

However, conservation aid usually arrives with strings attached to it, and this can generate resentment and can throw up some apparent anomalies, such as when external funding is readily available for moving elephants to Campfire areas but not to understocked safari areas, because donors do not want to be associated with sport hunting in the safari areas, despite the fact that elephants will be sport-hunted in both areas.[20] The difference is explained by donors wishing to placate their sponsors or memberships. In the North, safari areas are not funded because sport hunting is not favoured and not understood as a conservation method. Campfire, in contrast, though reliant on sport hunting, can be presented to the public as rural development mixed with conservation, which is favoured. For example, WWF-UK recently shifted from concentration on single species conservation, and Campfire was targeted for WWF funding precisely because it contained a development component and was perceived as a conservation project that incorporated social issues, rather than treating conservation as an exclusively ecological issue.[21] In contrast, smaller charities tend to concentrate on specific or single species projects. The Sebakwe Black Rhino Trust raises funds solely for rhino conservation in the Midlands Conservancy in Zimbabwe, which is possible, in part, because it is not a membership-based charity, but instead relies on fund-raising activities.[22] Predictably, most NGOs fund projects or areas that fit best with their own mission or with the demands of their memberships.

Another important distinction between NGOs is trade versus field organisations. Trade organisations tend to be more autonomous than small field organisations, which rely on a good working relationship with national governments. Trade organisations seek to control or prohibit trade in animal products, whereas field organisations seek to protect animals in their countries of origin. For example, the Environmental Investigation Agency (EIA) ran public awareness campaigns on the ivory and rhino horn trade. In the case of rhino horn trading, the campaigns have not only concentrated on range states, but have been extended to criticise the practices of Asian countries, which are the source of demand for horn, and so ultimately the source of poaching in range states (Pearce, 1991: 68–72; EIA

---

[20] Interview with Charl Grobbelaar, Chief Executive of the Zimbabwe Hunters Association, 12.2.95, Harare.
[21] *Herald* 24.7.87 'Funding Priority to Zimbabwe and Zambia Wildlife'; pers. comm. Peggy Allcott, WWF-UK Africa Section, 15.11.94, Godalming, Surrey.
[22] Interview with John Gripper, Sebakwe Black Rhino Trust, 14.10.94, Ascot-Under-Wychwood.

et al., 1993; Currey and Moore, 1994). However, EIA does not provide funding for field conservation projects. Other charities and NGOs do not involve themselves in campaigning but do specifically fund field projects. For example, Tusk recognises the value of campaigning organisations but is itself very much a field operator: its director, Charles Mayhew, suggested that campaigning organisations require salaried staff and a lot of time to effectively lobby other organisations and governments, while Tusk devotes its funds to field projects.[23] In addition, some NGOs do not lobby or campaign, because it could jeopardise their field projects. John Gripper, of the Sebakwe Black Rhino Trust, commented that, if the Trust were to trumpet the failings of the Parks Department in Zimbabwe, as EIA does, it would be made very unwelcome. If the Trust were unable to cooperate with the Parks Department over black rhino conservation, then its funding would be wasted.[24] To a large extent, EIA has been able to denounce Zimbabwe's actions without any significant negative effect on its programmes and campaigns, whereas field organisations are tied by the need to retain government support.

However, other small field conservation organisations have been involved in more clandestine activities. Rhino Rescue operates only in Zimbabwe and cooperates with some other NGOs. They are involved in rhino translocation to commercial farms and subsequent protection of the black rhino. One interviewee commented that it was recognised that Rhino Rescue did some great work for rhinos, but that its sources of funding were a mystery, although it was closely linked with Prince Bernhard of the Netherlands and David Stirling, founder of the British SAS (Special Air Service). In addition, their treatment of suspected poachers was open to question, with many being shot on sight (Pearce, 1991: 91–2). The main agenda of Rhino Rescue is to prevent poaching and, as part of this, in 1994, through its members Count Coreth and Colonel Dyke, Rhino Rescue assisted the Midlands Black Rhino Conservancy with funding for training guards and the creation of a rapid reaction force.[25] In Britain, Rhino Rescue belongs to the UK Rhino Group, an umbrella organisation, which includes the Sebakwe Black Rhino Trust, EIA, WWF, London Zoo, Care for the Wild and the David Shepherd Foundation, amongst others.[26]

In spite of numerous differences between international NGOs, they do face some common difficulties. Some can campaign and operate with little regard to the reaction of the Zimbabwean Government. On the other hand, smaller NGOs that prefer to fund field-based projects have found

---

[23] Interview with Charles Mayhew, Director, Tusk, 14.11.94, London.
[24] Interview with John Gripper, Sebakwe Black Rhino Trust, 14.10.94, Ascot-Under-Wychwood.
[25] *Sebakwe Black Rhino Trust News* April 1994.
[26] Interview with John Gripper, Sebakwe Black Rhino Trust, 14.10.94, Ascot-Under-Wychwood; see also Pearce (1991: 91–2.)

confrontational campaigning styles undermine their effectiveness. Larger international NGOs have a greater freedom to criticise government while remaining active within Zimbabwe, because of their access to funds for conservation.

Environmental campaigning

International NGOs have to communicate their conservation philosophy and their project commitments to the public in order to ensure that they continue to receive donations. In the 1980s, fund-raising was commercialised and became an increasingly professional business, with international conservation organisations recognising the potential of corporate fund-raising, pay-related giving and special events. However, donations from the public and membership subscriptions are still the main source of funding for large international NGOs. For example, in Britain, in 1994, two million people belonged to the main environmental and wildlife campaign groups, and this was more than the total memberships of all the political parties.[27] For some campaign-based NGOs, communication with the public and raising revenues is their *raison d'être*, and, as a result, NGOs tend to be skilful media manipulators. For example, Greenpeace relies on publicity to draw attention to environmental problems and the activities of the organisation and, with a publicity budget of US$1.5 million (one fifth of the total budget of the organisation) in 1995, it produced its own press releases and footage for television.[28] Similarly, EIA produces footage for television and places adverts for its campaigns in newspapers (Thornton and Currey, 1991). The ability to utilise the international media and effective campaigning styles meant that NGOs learnt that states and multinationals have a soft underbelly, thereby giving NGOs a degree of lobbying power over them.[29] Effective campaigning is often reliant on simple messages or concepts, which can be encapsulated in slogans that are easily understood, but issues such as sustainability are not so easily communicated with snappy slogans.

WWF has encountered major problems in communicating its policies and ideas of good conservation practice to the public, but first it is useful to make a distinction between WWF-International and the national organisations around the globe. WWF-International has been criticised by external commentators and its own national chapters. WWF was established in 1967, when it exploded into public consciousness with media campaigns in the tabloid press to save rhinos and elephants. Initially, WWF was set up to raise funds for scientific projects carried out

---

[27] *Independent* 17.11.94 'What On Earth Do They Do Now?'; *Independent* 30.1.96 'A Professional Tug at the Heart'; de Waal (1997).

[28] *Independent* 5.9.95 'Covering Itself in Glory'.

[29] *Economist* 20.7.96 'The Fun of Being a Multinational', pp. 63–4.

by the International Union for the Conservation of Nature (IUCN), but in the 1980s WWF split from IUCN to pursue its own projects. WWF-International has been criticised for its remoteness from national chapters and its sources of funding. The very public and acrimonious split with WWF-USA in the early 1990s demonstrated the depth of feeling about what national organisations viewed as wasting of resources, the top-heavy nature of the organisation and the feeling that WWF-International was out of touch with conservation in the field (the bitter split was healed in 1992 with an agreement between WWF-USA and WWF-International to work together, although WWF-USA retained its autonomy) (Pearce, 1991: 69–105; Bonner, 1993: 72).

The use of resources by WWF was questioned by other conservation organisations, since many WWF officials have a high public profile, spend much time jetting between conferences and are highly paid. While many of WWF-International's projects are in developing countries, its headquarters remain in Switzerland, and this has reinforced the perception of WWF-International as divorced from the reality of conservation around the world. Similarly, African chapters of the organisation complained of a lack of consultation over position statements and funding for projects within Africa and levelled charges at WWF-International of neocolonial attitudes towards national chapters and of leading WWF figures behaving in an egocentric manner (Bonner, 1993: 74–7). In many ways, WWF-International is a patrician organisation; for example, in his preface to the 1970 WWF publication *Wildlife in Crisis,* Prince Philip, President of the organisation, stated that Europeans never restricted their interest in conservation to home countries and that colonial administrations had an excellent conservation record (WWF, 1970: 17). This view was rejected by a number of commentators, who suggested that colonial administrations did much to upset the natural environment and were more interested in killing African wildlife than in saving it (Kjekshus, 1977).

WWF as a whole has been subject to criticism for its policies and for the actions of representatives. WWF-UK is often cited as having the biggest problems with its constituency, an example being that WWF-UK supports sport hunting but cannot publicly promote it. The British-produced 'Cook Report' filmed an exposé of WWF-Zimbabwe's involvement in sport hunting in July 1990. In it, a journalist posed as a safari hunter and filmed the hunt, linking it to WWF's Multispecies Animal Production Systems (MAPS) programme, and the film was then used as a stick to beat WWF, claiming that the organisation was directly involved.[30] WWF-UK has been partially involved in sport hunting and has felt unable to communicate the potential benefits for conservation to

[30] Interview David Cumming, Director, WWF-Zimbabwe, 22.2.95, Harare; see also Bonner (1993: 263).

its membership, because it feared that its members would leave.[31] WWF-UK also drew up a position paper for answering questions on sport hunting in the media. It was a confidential document and attempted to explain sport hunting as a conservation method (WWF, undated). WWF as a whole is in a difficult position, since it assists in field projects based on sustainable utilisation but is afraid to discuss it openly as a policy option in public forums.

The figures involved in WWF-International have also been criticised. The profile of the funders raised doubts about the organisation's ability to remain true to its principles and membership while also satisfying funders. Conflicts of interest between companies allied to WWF funders and conservation have arisen and it is notable that WWF does not engage in the same level of public denunciations of companies, such as Shell Oil and logging companies, as Greenpeace does.[32] The list of trustees includes numerous directors of multinational corporations, whose commitment to global environmental protection is questionable. Trustees also often have common business interests, such as Sir Kenneth Kleinwort of the Kleinwort Benson Group, who is a trustee alongside Sir Arthur Norman, who served as Director of Kleinwort Benson from 1985 to 1988. Clearly, the list of WWF-International funders reflects its patrician nature.

One of the most controversial figures was the South African businessman Anton Rupert, who joined the Executive Board of WWF-International in 1968. Rupert was the Managing Director of the Rembrandt Group of tobacco companies; he is also the Director of Rothmans Tobacco Holdings and is a member of the South African Chemical Institute. Tobacco growing is known to be damaging to the surrounding environment and the commitment of chemical manufacturers to conservation is highly questionable. He was also the President of WWF-South Africa, which was then named the Southern African Nature Foundation (Peace Parks Foundation, [1997?]). Once Rupert joined WWF-International, he quickly came up with the idea for the 1001 Club, which was to consist of some of the wealthiest people around the globe. In return for a donation of US$10,000 per year, members' names were kept secret and they joined a very exclusive club. The 1001 Club was a welcome opportunity for a number of South African businessmen to mix internationally at a time when sanctions were in effect and South Africa was politically and economically isolated. Similarly, the membership list indicates that 1001 Club financiers included President Mobutu of Zaïre, Agha Hasan Abedi of the Bank of Credit and Commerce International, which dramatically collapsed in 1991, and Daniel K. Ludwig whose company has destroyed great swathes of South American rain forests (Bonner, 1993: 62–72). In the case of the ivory trade, WWF-

---

[31] Interview with Stewart Chapman, WWF-UK CITES Adviser, 15.11.95, Godalming, Surrey.
[32] *Independent* 5.9.95 'Covering Itself in Glory'.

International had the dual problem of satisfying its funders and its membership base. In general, its members were influenced by campaigns to ban the ivory trade. However, WWF-International argued against a total ban for some time, and it could be suggested that this was to appease funders in the 1001 Club, who were prominent South Africans and were against a ban. The position of NGOs is clearly more complex than their campaigns suggest. Conservation campaigns and funding of projects appear to follow a coherent position, mostly a preservationist one, but actual policy may differ among various parts of the organisation. Nevertheless, the campaigns indicate a commitment to articulating a particular political ideology of conservation, thereby contradicting the argument that the organisations save animals through the application of socially, politically and economically neutral scientific management principles.

Concepts of animal rights have had a significant impact on the conservation policies promoted by international NGOs and Western governments. Deep-green commitments to preservation have been promoted by international NGOs, which, in turn, exercise such power over the conservation policies of African states through funding that NGOs such as the Humane Society and EIA are capable of buying influence in Africa.[33] The animal rights lobby has a disproportionate amount of influence compared with its membership in the USA. For example, the Humane Society of America recently produced a report on the failure of the rhino dehorning programme, concluding that Zimbabwe should halt what it called an inhumane technique. In response, Professor Marshall Murphree, director of the National Parks Board, called the report 'scandalous, inaccurate and a prostitution of science'.[34] Reports such as this do have an effect on conservation, because they can be used to persuade external donors (NGOs or governments) not to contribute to certain conservation practices in Zimbabwe.

Appeals to notions of animal rights permeate debates on what constitutes good conservation practice and the ethics of management plans. The belief in animals as sentient beings with a right to existence separate from any value to human beings is reiterated by a number of organisations and underlies strict preservationist policies. For Zimbabwe, this idea is translated into disapproval for some forms of sustainable use of wildlife. According to Jon Hutton, the debate over sustainable use has been emotionalised by animal rights organisations, which now define policies in terms of lethal or non-lethal use of wildlife.[35] Animal rights organisations do adhere to preservationism, since they are interested in saving every single animal. Other conservationists argue that, in the long term, this may operate against the interests of the species as a whole. For

---

[33] *Economist* 20.4.96 'Killing to be Kind', pp. 100–1.
[34] *Sebakwe Black Rhino Trust Newsletter* April 1994.
[35] Interview with Jon Hutton, Director of Projects, Africa Resources Trust, 13.3.95, Harare.

example, allowing only non-consumptive use of elephants in Zimbabwe could result in overpopulation in some areas, thereby leading to starvation and the death of whole herds. Zimbabwe's policy philosophy favours a utilitarian approach, where the trophy fee from one sport-hunted elephant can pay for conservation for many other elephants (Ministry of Environment and Tourism, 1992).

Since arguments over the rights of animals are related to political ideologies, animal welfare organisations are in many ways political organisations. Their activities revolve around lobbying governments and other institutions and, as such, they address moral and ethical issues, rather than engaging in field conservation. This partly reflects a split between field organisations and those concerned with the moral dimensions. Of course, the Parks Department has criticised external organisations that denounced what the Department saw as realistic conservation policies on the grounds of animal rights. However, the individual objectives of single NGOs have the capacity to jeopardise this policy. In 1995, the US-based Humane Society informed the Zimbabwean Government that it was filing a law suit against the US Government to prevent sport-hunted ivory entering the USA from Tanzania, and a ruling on this could be extended to include sport-hunted ivory from Zimbabwe (Rowell, 1997: 34–6). Animal rights have the potential to influence the conservation policies of individual states, and this is most starkly represented by the international debates over culling.

The Zimbabwean Government has consistently claimed that the country has too many elephants for the land and the society to support. Estimates of elephant numbers vary according to the argument the counting organisation accepts. The Parks Department estimated that, in 1989, Zimbabwe had 51,700 elephants (Martin et al., 1989: 18); and official figures reported in the Zimbabwean media put it as high as 80,000.[36] However, other estimates given by the Parks Department and repeated by NGOs put the population between 65,000 and 70,000 (EIA, 1992: 25). The official figures are disputed by anti-culling organisations, which suggest that Zimbabwe exaggerated its elephant population in order to strengthen its position on ivory trading, sport hunting and culling. Critics argue that the rapid rise in the elephant population cannot be explained by natural reproduction, because it has risen too quickly, and so it must be due to migration of elephants to the relative safety of Zimbabwe from troubled and heavily poached areas in Mozambique, Zambia and possibly Angola (Ivory Trade Review Group, 1989: 6; Douglas-Hamilton et al., 1992: 85). In addition, NGOs suggested that Zimbabwe was guilty of double counting their elephants by counting them once and then allowing them to migrate across the border into Botswana, Mozambique and Zambia, only to be re-counted once they returned with changing seasons (EIA, 1992: 25). Both sets of interest

[36] *Herald* 21.7.95 'Elephant Count Underway in Southern Africa'.

groups have a reason to adhere to the argument that they present. The Parks Department could overestimate elephant numbers so that they have a stronger case to cull them and sell their products on the international market. The critics may wish to underestimate the population to undermine the Zimbabwean position. In order to resolve this problem of double counting and overestimates, the European Union sponsored a new elephant count in 1995, to be carried out simultaneously in Malawi, Zimbabwe, Botswana, South Africa and Zambia and intended to standardise elephant censuses in the southern African region.[37] Part of the reason for Zimbabwe's position is that an industry has grown up around culling. The hides, meat and ivory can be sold for profit within Zimbabwe. The Parks Department has stated that, if it were allowed under CITES, it would sell the products of culling on the international market and that such sales would increase the economic value of culling. Critics argued that this could result in elephants being culled for economic reasons rather than ecological ones (Care for the Wild, [1992?]: 28). However, the Parks Department argues that culling also provides an important social good because meat from culling impala and elephants was made available to local people during the 1991–2 drought and, without such a programme, the government would have been acting in an inhumane manner towards its people.

Culling is a highly contentious issue for conservationists, because, at its heart, culling involves the question of whether nature is able to regulate itself or whether management policies should be devised. Conservationists have been divided over whether to interfere in elephant populations through culling. The choice is often presented by NGOs as between culling and allowing elephants to die off naturally as a result of localised overpopulation (Chadwick, 1992: 103–6). If elephants become too numerous, they can destroy their own habitats, and this results in a slow death from malnutrition and starvation. In addition, it can also force other species, such as rhino and giraffe, to move elsewhere in search of food, or they will face the same fate as the elephants.

The Parks Department insists that there are too many elephants in certain areas of Zimbabwe and so, from time to time, they are culled to control population and elephant damage to vegetation. Culling has had a relatively long history in Zimbabwe, with the first major cull in Hwange National Park in 1965, and then in Mana Pools in 1969, 1970 and 1972. These culls were carried out because it was believed that the loss of mature trees, due to damage caused by elephants, was resulting in the decline of the black rhino in those areas. As a result, between 1965 and 1989, approximately 44,500 elephants were culled (Martin et al., 1989: 5). More recently, due to the prevailing drought conditions of the 1990s, culls have been carried out in Gonarezhou National Park.[38] Zimbabwe is

---

[37] *Herald* 21.7.95 'Elephant Count Underway in Southern Africa'.
[38] *Herald* 6.7.92 'Thousands of Animals to be Culled in Gonarezhou'.

very sensitive about its culling programme; conservationists argue that it is done with the ethical and practical goal in mind of maintaining biodiversity (Zimbabwe Trust, 1992). As a result of sensitivity, the Parks Department has modified its culling practices in line with new environmental information. For example, the practice of shooting older elephants and babies in culls has been replaced by culling whole families (Pinchin, 1993). This is because more recent research suggests that such practices removed older elephants, who held critical learned knowledge of ancient routes to water for use in time of drought. Originally, it was thought that it was best to kill elephants with no breeding potential, as it would not affect the ability of elephants to survive in the long term.

Culling is highly unpopular within the international community and is used in attempts to politically discredit Zimbabwe's conservation record at CITES. Zimbabwe is currently attempting to expand its tourist industry, and most of its visitors arrive from the industrialised nations. The Zimbabwe government is keenly aware that bad publicity surrounding culling can stifle tourism development almost as soon as it has begun. Opponents of culling present it as cruel and disruptive to remaining herds in the area. Those involved in culls have reported how groups of elephants far distant from a cull site will immediately turn and move towards the commotion. Recent research has revealed that elephants can communicate for up to 10 kilometres with low frequency sounds transmitted through the ground, and this is thought to explain how remaining herds quickly congregate at the cull site.[39] While most conservationists, those who favour culling as well as those who do not, agree that elephants grieve over the loss of members of their family, critics of culling suggest it disrupts family life in the long term and results in behavioural problems in young elephants. Some NGOs, such as EIA, argue that culling is always wrong and is cruel, and that there is no scientific evidence of the need to cull in Zimbabwe. Another criticism is that often vegetation is damaged before a cull takes place, so the only reason to cull is to preserve the appearance of remaining vegetation for tourists (EIA, 1992: 25).

One possible solution to culling is elephant translocation, and Zimbabwe is a world leader in this field. Translocation involves moving elephants from areas of overabundance to areas where there are not enough elephants. It was in the 1991–2 drought that the first major translocation effort was undertaken, when elephants were moved from Gonarezhou National Park to the Save Valley and Bubiana private conservancies, and some of the elephants were also sold to private game ranches in South Africa.[40] The problem with translocation is that it is expensive. In general, it is carried out by a private company, run by an ex-

---

[39] *Independent* 26.2.96 'When an Elephant Gets Emotional'; Care for the Wild ([1992?]: 31).
[40] *Herald* 2.4.92 'Save Our Game Campaign in the Lowveld'; *Sunday Mail* 21.3.93 '600 Animals Translocated'.

Parks employee, Clem Coetsee, through his company, Wildlife Management Services, and in the 1991–2 translocation exercise it was estimated that the move would cost £300 per elephant.[41] A number of NGOs, local and international, have been involved in providing the necessary finance. For example, Tusk was active in fund-raising for Operation Oasis in 1991–2. Fund-raising events and an emergency appeal resulted in massive response from the public. The German government committed DM1,750,000, which was channelled through GTZ (the German government development agency).[42] In addition, Care for the Wild assisted the local NGO, Save Our Wildlife Heritage run by Penny Havnar, by providing funds.[43] Those who receive the elephants often pay market prices for them, and in 1993 the Parks Department expected to earn Z$1.2 million from sales of 150 elephants to Bophuthatswana for Z$3000 each. Such sales can contribute to the costs of capturing and translocating the elephants.[44] Translocation is promoted as a viable alternative to culling by organisations such as Care for the Wild, Tusk and the International Fund for Animal Welfare (IFAW).

However, some local NGOs were concerned by the cost of translocation and suggested that, if international NGOs wanted Zimbabwe to turn to translocation and abandon culling, they should pay for it. Nevertheless, within and outside Zimbabwe, translocation is widely publicised and presented as proof that culling is no longer necessary (Care for the Wild, [1992?]). The promotion of translocation as a real alternative has meant that the Parks Department has been less able to ignore international pressure to stop culling. In this way, the Parks Department has been deprived of the argument that there is no other way to reduce elephant numbers. The strength of animal rights organisations is such that, due to international pressure, Zimbabwe abandoned a planned cull in Hwange in 1994 and allowed the elephants to walk across the border into Zambia.[45] It is clear that international NGOs have a significant impact on policy decisions in Zimbabwe. They are able to exert influence through campaigning and via the large funds that they have at their disposal.

International donors

To a large extent, government-to-government aid and funds that are offered by international financial institutions, such as the World Bank, can also induce certain policy outcomes. A number of overseas governments and their development agencies provide funding for conservation

---

[41] Care for the Wild News Summer 1994 'Operation Elevacuation'.
[42] Tusk Trustees Report 1992–3.
[43] *Care for the Wild News* Summer 1994 'Operation Elevacuation'.
[44] *Herald* 27.3.93 'National Parks to Earn $1.2 Million From Elephants'.
[45] *WWF News* Summer 1994.

projects in Zimbabwe. As with NGOs, their choices of funding often follow the same pattern, since donors channel money through NGOs, in the belief that they are the best means of getting aid to the target groups (Meyer, 1992). This means that a great deal of aid money completely bypasses central treasuries, much to the annoyance of governments. There has been very little research on the effectiveness of NGOs, but the assumption that they are often preferable to central governments continues (Vivian, 1994).

In the mid-1990s, the British Overseas Development Administration (ODA) (recently renamed the Department for International Development (DFID)) stopped directly funding the Parks Department in Zimbabwe, because of its internal power struggle. The ODA did not have any confidence in the Department or the government, believing that funds for conservation would be used to fill gaps elsewhere. This certainly occurred when USAID granted Z\$3.5 million to alleviate chronic water shortages in Hwange National Park, and it was later discovered that Z\$800,000 of that grant was diverted to pay for general administration costs, including air fares for senior Parks staff.[46] As a result, the ODA stated that it required some restructuring of Parks and a commitment from the central treasury to pass on aid money for conservation to the Parks Department.[47] It was hoped that such pressure would induce policy changes and restructuring. In this way, donors can effectively buy influence within a state. This was a clear case of political conditionality, in line with the ODA's stated concern with good governance. Despite ODA concerns with the Parks Department, Britain is Zimbabwe's biggest single donor and bilateral aid was increased in 1995 to £50 million to provide financial support for World Bank-backed reforms.[48]

However, governments do sometimes provide finance directly for government departments in Zimbabwe. It is widely accepted that the Parks Department is underfunded and is often unable to maintain basic services, so finance is occasionally provided for specific, usually high profile, projects, which the Parks Department cannot afford to carry out. For example, in 1994, the German Government provided Z\$103,000 for water pumps in Hwange National Park, because many animals were in danger of dying from thirst and a number of diesel pumps had either failed or were without fuel to power them.[49] The Parks Department states in its own policy documents that external funding is sought for certain conservation projects. For example, black rhino conservation is extremely expensive, requiring an annual budget of US\$20 million and most of this funding must be from external sources (DNPWLM, 1992c).

[46] *Daily Gazette* 3.11.94 'Aid Funds Diverted'.
[47] Interview with Caroline Plastow, Third Secretary (Aid), British High Commission, 27.3.95, Harare.
[48] *Financial Gazette* 2.3.95 'Britain Shifts Aid Focus to Balanced Budget'.
[49] *Development Dialogue* Dec. 1994–Jan. 1995 'Germany Gives Z\$103,000 for Water Pumps'.

USAID differs markedly from its British counterpart, the ODA, in that it does provide finance directly to the Parks Department. This is viewed as the best means of inducing change in the Department, so the Embassy is responsible for monitoring the political situation in the Department, while USAID provides the 'carrot' for change in the form of a cheque book for conservation funding.[50] This carrot and stick approach is also used by the World Bank to induce policy changes and to modify the internal workings of the Parks Department.

Donors have fairly rigid guidelines, which govern decisions about funding, and, in the case of Campfire, these guidelines coincide with the interests of the Parks Department. However, decisions to fund some areas and not others can directly frustrate the broader objectives of the Department. The ODA provided funding for Campfire through Zimbabwe Trust and WWF under a joint funding scheme, which means that the NGOs must match the ODA contribution dollar for dollar.[51] Campfire has been favoured by the ODA and other donors for many of the same reasons it is funded by NGOs. Campfire provides an example of conservation with a strong social welfare or rural development component. This is important for donors, which need to be viewed as caring about people and animals, rather than simply ploughing money into animal welfare. As with the Parks Department, external donors can use Campfire's commitment to conservation coupled with rural development to provide a political justification for supporting conservation for their domestic constituencies and on the international stage.

A number of Parks Department representatives stated that Campfire did not need donor funding, since it was self sufficient, even producing large profits, from an early stage. Yet, as the aid was offered, the Parks Department and other NGOs involved in Campfire accepted the extra finance. George Pangeti of the Parks Department suggested that Campfire receives so much donor support because NGOs and donors like to be associated with a success story and Campfire is a 'sexy project'.[52] It was impossible to obtain figures on donor funding for Campfire, because they were held confidentially by the Ministry of Environment and Tourism and the Ministry of Finance. However, one source suggested that USAID alone had provided US$1 million.[53] Yet, in a sense, USAID hedged its bets over Campfire, since USAID was a major donor to Campfire but was also funding traditional and coercive forms of anti-poaching, which conflicted with Campfire objectives. Consequently, in the domestic and international political arenas, USAID could pronounce anti-poaching to be its success story if Campfire failed in the early stages (King, 1994: 32–3).

---

[50] Interview with Stephen T. Bracken, Second Secretary, US Embassy, 21.3.95, Harare.

[51] Interview with Caroline Plastow, Third Secretary (Aid), British High Commission, 27.3.95, Harare.

[52] Interview with George Pangeti, Deputy Director, DNPWLM, 21.7.95, Harare.

[53] Interview with Stephen T. Bracken, Second Secretary, US Embassy, 21.3.95, Harare.

The World Bank has expressed increasing interest in funding wildlife ventures around the globe and this has especially been the case since the 1992 Rio Summit. The United Nations Conference on Environment and Development (UNCED) agreed to the establishment of a fund to provide finance for environmental projects. The Global Environmental Facility (GEF) was the subject of much debate. In general, developing states wanted GEF to be controlled by the United Nations Environment Programme (UNEP), while the Western states were keen to see it in World Bank hands. After much wrangling, GEF was placed under the auspices of the World Bank, and funds have certainly been offered for conservation projects of a particular type. International financial institutions, such as the World Bank, follow the same patterns as NGOs, because, in general, they provide finance for projects that intersect with their underlying policy principles and stated areas of interest (see World Bank/GEF, [1997a?, b?]). Finance from GEF has been offered to Zimbabwe for the decentralisation and commercialisation of a government department and the establishment of a tourist-centred superpark. The World Bank has indicated interest in funding such a superpark. A superpark can be defined as a wildlife conservation area that follows ecological boundaries rather than being confined to the borders of a nation-state (Peace Parks Foundation, [1997?]). The artificial colonial boundaries in sub-Saharan Africa do not correspond to bioregions, and as a result ecosystems were split between a number of states. The World Bank has carried out feasibility studies on three areas, Dongola (South Africa, Zimbabwe and Botswana), Gaza (Zimbabwe, Mozambique and South Africa) and Chimanimani (Mozambique and Zimbabwe).[54]

The superpark would undoubtedly provide conservation benefits, such as reopening old wildlife migration routes and increasing the genetic diversity within species. However, it is the economic potential of the area that the World Bank has been keen to promote. Its main advantage is perceived as economic. In general, tourism is poorly developed in the south-east of Zimbabwe, and the main tourism centres for international visitors are concentrated around the Victoria Falls and Hwange area. The creation of a superpark in the south of the country has been presented as a means of stimulating an economically depressed area with few industries and little potential for anything else but wildlife development. It is intended to take advantage of an expected growth in regional tourism by Western visitors and the newly welcome South Africans.[55] The superpark would provide tourists with the largest wildlife conservation area in Africa, and it is intended to become part of a southern Africa tourism circuit for international visitors and part of a

---

[54] *Financial Gazette* 16.2.95 'Cross Border Parks Shape Up'.
[55] Interview with Tarirai T. Musonza, Research and Planning Officer, Zimbabwe Tourism Development Corporation (ZTDC), 23.3.95, Harare; *Herald* 4.5.95 'SA Citizens to Get Free Visas'; Peace Parks Foundation, [1997?].

South African and south-eastern Zimbabwe circuit for regional tourists.[56] In addition, the recently established Save Valley Conservancy, which is a private wildlife ranching concern, has been keen to promote the superpark. Clive Stockil, chair of the Conservancy, stated that the Conservancy could benefit tremendously by eventually becoming part of the superpark.[57] The superpark has been perceived as a great marketing opportunity, promoted to international clients as part of the largest wildlife area in Africa. Such a link would provide visitors from America and Europe with the opportunity to take vehicle safaris through the Kruger National Park and into Zimbabwe (Duffy, 1997).

In accordance with prevailing donor wisdom on environment and development initiatives, the World Bank has emphasised the inclusion of local communities as a condition for its support. One of the criticisms of the superpark concept is that it follows a traditional pattern of national park development that has clearly failed across the continent (Ellenberg et al., 1993: 31). The exclusion of local people from wildlife conservation areas has led to incursions beyond park boundaries by local communities. Consequently, current park plans have included provisions for allowing adjacent communities to share directly in the benefits from them. However, funders and park planners have been vague about precisely how local communities can benefit beyond obtaining employment in tourist facilities (Ellenberg et al., 1993; Sguazzin, 1995: 1).

Despite the economic potential of the development of a superpark, the central government in Zimbabwe has proved a recalcitrant partner in the scheme. Caesar Chidawanyika of the World Bank stated that South Africa and Mozambique had been very keen, while the Zimbabwean government had 'yet to be sold on the idea'.[58] The reaction of ZANU-PF to the superpark can be explained as the result of a number of complex factors. The government has used the loss of sovereignty argument to delay the scheme, because the park would involve agreeing to joint authority over an area of Zimbabwean territory, and the dominance of international donors in the project led to frustration in the government with donors perceived as being in control of policy-making. The central government was keen to point out that it would not bow to such imperial tendencies, while clearly following a national plan of World Bank-inspired reforms anyway.[59] However, the sovereignty argument has proved to be a useful screen behind which to hide other pressing concerns about the creation of a superpark.

---

[56] Interview with Caesar Chidawanyika, Senior Programme Officer, World Bank, 3.5.96, Harare.
[57] Interview with Clive Stockil, Chair of the Save Valley Conservancy, 13.5.96, Chiredzi.
[58] Interview with Caesar Chidawanyika, Senior Programme Officer, World Bank, 3.5.96, Harare.
[59] *Herald* 2.2.95 'Donors Won't Dictate Policies Says Mugabe'.

To explain government resistance to the plans, it is necessary to place the superpark plans in the wider political context of conservation in Zimbabwe. The superpark has been promoted as an extension of the World Bank's plans for commercialisation of the Parks Department. Supporters included the former Minister of Environment and Tourism, Herbert Murerwa, a number of senior Parks officials such as Rowan Martin, George Pangeti and Willie Nduku, and the World Bank. Supporters suggested that restructuring would wrest control of the Parks Department out of the vagaries of central government, allow it to retain the revenue it generates and free the Department from corruption. Conservationists in Zimbabwe, while concerned that wildlife should not be in private hands, were aware that there was a need for change in the Department.[60] However, these proposals were not approved until 1996, and it was widely believed that they were swallowed by bureaucracy and sabotaged by internal ZANU-PF politics. A number of senior Parks officials were opposed to decentralisation and commercialisation because they feared a loss of control over Parks Department decisions and, most importantly, resources. One interviewee suggested that commercialisation and the superpark were opposed because the government did not view competition as a means to improve the Parks Department's operations. This was part of the Department's inability to reconcile its role as a regulatory body with its need to be a commercial operation dealing with the demands of the tourism industry.[61]

The superpark proposals and commercialisation plans also represented a threat to the central treasury, since the scheme would benefit private operators and the Parks Department, which would obtain revenue that would previously have been collected by the treasury. In addition, decentralisation of conservation in community development initiatives has been frustrated by the central treasury, because it perceives such schemes as a threat to its revenue. The Ministry of Finance was reported to be attempting to block the restructuring of the Parks Department, and currently the Parks and Wildlife Estate is a major income earner for the central treasury.[62]

A final reason for government recalcitrance is the history of poaching and the security situation in the area of the proposed superpark. The Mozambican area of the proposed park still contains people who will have to be disarmed and moved. One source also contended that the park was being frustrated by elements in Zimbabwe that were concerned that an increase in the tourist traffic in the area called crooks' corner (where the borders of Zimbabwe, South Africa and Mozambique meet) would

---

[60] *Herald* 23.3.94 'Time For Action on Corruption'; *Financial Gazette* 24.3.94 'Department Replete With Corruption'.
[61] Interview with Raoul Du Toit, WWF/DNPWLM, 7.5.96, Harare.
[62] Interview with Stephen T. Bracken, Second Secretary, US Embassy, 21.3.95, Harare; see also Duffy (1997).

deny them the continued opportunity of poaching (Ellis, 1994; Duffy, 1997, 1999). The superpark constitutes a definite challenge to the traditional boundaries and powers of the nation-state. While conservation policies are presented as the realm of an uncontested pure science or scientific management principles, it is clearly a highly political issue, which reveals wider difficulties within national and regional politics.

In conclusion, local and international conservation NGOs have a dual role to play in Zimbabwe. Local and international organisations have formed collaborative links with the Zimbabwean state for mutual benefit, while, on the other hand, there have been significant conflicts between the state and NGOs. In general, local organisations have played a more supportive and collaborative role, but, while they tend to be supporters of sustainable utilisation anyway, numerous NGOs have found cooperation more productive than conflict. The Parks Department has benefited in terms of the publicity and financial support that local NGOs have provided, and equally local NGOs have made use of their links with the government. International NGOs and donors have had a more ambiguous role to play. Numerous international organisations, such as Sebakwe Black Rhino Trust and Rhino Rescue, have also found a cooperative stance to be mutually beneficial to them and the Zimbabwean state. However, other organisations, such as WWF, have had a more ambiguous role as supporters and critics of Zimbabwe's conservation policy. The support gained from these local and international organisations has had an influence on policy in terms of providing publicity, equipment and funding, and their support has assisted the Zimbabwean Government in continuing with its policy of sustainable utilisation.

However, NGOs that have preferred a more confrontational relationship with the Zimbabwean Parks Department have also had an influence. It is for this reason that donors have been accused of ecological imperialism, where they exercise control over conservation policies within other states without really having any responsibility over what the outcomes of those policies might be. Animal rights-orientated campaigning organisations have influenced Zimbabwe's policy choices through publicity, denunciation and funding campaigns against utilisation, and this was nowhere more apparent than over the culling issue. The Zimbabwean authorities have become sensitised to international opinion about culling practices through campaigns run by NGOs, such as EIA and Care for the Wild. Care for the Wild, in particular, influenced the Parks Department's switch to translocation as an alternative to culling, because it provided funding for moving elephants. In this sense, the Parks Department and Care for the Wild had shared interests, despite the their adherence to opposing environmental philosophies. It is clear that donors and NGOs have indeed been able to buy influence in Zimbabwe and have competed with other interest groups that have aimed to exercise a level of control or influence over the direction and operation of conservation policy. NGOs, both national and international, are an important force in wildlife policy, and form formidable alliances to support or attack the policy of sustainable utilisation. In doing this, NGOs assert their own political agenda, which also assists in justifying their actions to their own constituencies or memberships and deflecting attacks from their opponents and critics.

7

The International of the Rhino Horn
Politics & Ivory Trade

Trading in wildlife products was one of the most emotive issues in conservation and, during the 1980s and 1990s, this was indicated in the high-profile media campaigns by conservation organisations aimed at halting the trade. However, the campaigns surrounding saving elephants and rhinos failed to examine the complex nature of the politics of the trade in ivory and rhino horn. In particular, the trade in wildlife products operates within an international legislative and ideological framework that is often at odds with national policy agendas. This is nowhere more apparent than in the debates over the role of the Convention on International Trade in Endangered Species of Flora and Fauna (CITES).

The formation of global institutions to regulate the behaviour of states and non-state actors has been a central theme in international relations. Global conventions have been established in response to transboundary activities and problems where domestic legislation has proved inadequate, and so they are especially significant in the area of environmental management. They include attempts to regulate natural resources, such as the Law of the Sea, the International Whaling Commission and the Global Climate Convention. CITES recognises that the trade in wildlife and wildlife products can potentially undermine national conservation efforts, and this is clearly demonstrated by the trade in ivory, rhino horn and tiger bone. The trade in these products is global, because it reaches from range states in Africa and Asia, to middlemen trading states and finally to consumer states in the Far East, and such transnational trading, whether legal or illegal, clearly requires global management. In the 1960s and 1970s, domestic legislation in producing, trading and consuming states proved inadequate for controlling the ivory and rhino horn trade, and so CITES was drawn up in 1973. The focus of a number of global environmental agreements is prohibition, or the threat of prohibition, as a norm – that is, a standard rule to regulate the behaviour of parties. In the view of wildlife policy-makers in Zimbabwe, CITES has evolved from a regulatory regime to just such a prohibition regime with regard to certain wild species. The tensions over the decision to ban trade in ivory and rhino horn highlight the deep divisions between Zimbabwe and the international community over the operation of a global norm. This is especially significant in the light of recent moves towards a qualified lifting of the ivory ban for Zimbabwe, Botswana and Namibia, and in many ways this shift highlights the influence that producer states can have on international conventions.

The parties to the convention have a general common goal, in that CITES functions to prevent trade threatening species with extinction, and, in establishing a global agreement, parties to a convention must assent to a set of norms and ideas. In the case of CITES, obtaining a consensus on these norms has proved problematic, because sharp differences have arisen over specific wildlife issues and particular species. This is partly because of diverging conservation philosophies and the levels of bargaining power held by the parties. The industrialised states and East African states are in favour of preservationist methods, and together they form an alliance of interests, along with international non-governmental organisations (NGOs), which view total trade bans as the proper means of ensuring the conservation of elephants and rhinos. In contrast, the southern African states, including Zimbabwe, are committed to a policy of sustainable use and, along with a number of ivory- and rhino horn-consuming states, they constitute an alliance in favour of reopening a legal trade. This has been especially important for conservation policy in Zimbabwe. The influence of CITES as a regulatory regime has highlighted the fact that Zimbabwe's wildlife policy operates in a domestic and international context. This chapter will assess CITES, investigate the efficacy of the ban on rhino horn trading and demonstrate the difference between it and the ivory ban. It will analyse the controversy over the ban on the international trade in ivory, including an examination of pro- and anti-ban positions and the role of NGOs as moral actors within CITES. Finally, this chapter will assess the effectiveness of the ivory ban in the 1990s and recent Zimbabwean attempts to overturn it.

The Convention on International Trade in Endangered Species

International regimes have, in general, been formed by states seeking to control transboundary activities in order to guarantee benefits to dominant parties (Young, 1989). In the case of international environmental agreements, it is clear that certain norms, usually based on Western political ideologies, also have a critical role to play. These norms strictly control the conditions under which states can participate in and authorise certain activities (Nadelman, 1990: 479). CITES has proved to be the most significant international conservation agreement for the Parks Department in Zimbabwe. In most cases, CITES provides a framework for regulation of trade in wildlife and wildlife products, although in the cases of two of the most high profile wildlife products (ivory and rhino horn) CITES has operated as a prohibition regime. As a result, CITES has been an important determinant of domestic wildlife policy, since the trades in rhino horn and ivory have been the most contentious issues.

CITES is a highly developed international regime, designed to control and regulate trade in wildlife and its derivatives. The wildlife trade is a global business, which encompasses legal and illegal trade in animals and plants. Trade Records Analysis of Flora and Fauna in International Commerce (Traffic) is affiliated to the World Wildlife Fund (WWF)-International and the organisation is responsible for monitoring trade in wildlife, especially illegal trade. Traffic estimates that the value of the wildlife trade is second only to the drugs trade among illegally traded goods, but recognises that accurate figures are impossible to obtain, since much of the wildlife trade is illegal and unrecorded. Estimates of the value of the legal trade stand at around US$20 billion per annum, while Interpol suggests that the illegal trade is worth a further US$5 billion per annum (EIA, 1994: 3). The scale of the wildlife trade led to calls for an international body to regulate it in order to prevent overexploitation of certain species. The outcome of discussions was that CITES was drawn up in 1973 and came into force in 1975. Since then, the number of species it covers has increased and the number of member states has grown to over 120 (Lyster, 1985: 240). The basic regulatory tools and principles of CITES are set out in a system of appendices where an Appendix I listing constitutes a trade ban, Appendix II allows controlled and monitored trading and an Appendix III listing requires the trading state to notify CITES in order to prevent overexploitation. CITES does have provisions for species to be transferred between the appendices: for instance, to include, delete or transfer a species requires a two-thirds majority vote of the parties at the biennial conferences. In addition, the convention has provisions for a split listing of a species, which allows certain populations of a single species to be listed on a different appendix (Wijnstekers, 1994: 14–16).

Although the criteria for a CITES appendix listing are presented as scientifically determined, they are subject to stark political disagreement, because, in effect, an appendix listing operates as an international norm. In general, all populations of a single species are treated equally under an appendix listing, regardless of its local, national or regional status (with the rare exception of split listings). The system of appendices has proved the most contentious issue for Zimbabwe, because policy-makers in government institutions and NGOs argue that the basic norm that underlies most of the operation of CITES as an institution is the precautionary principle. This principle requires parties to demonstrate that trade is non-detrimental to the survival of the species being traded and, if there is any doubt, states are asked to cease all trade until non-detriment findings are proved (EIA, 1994: 5; O'Riordan and Jordan, 1995). It is the distinction between species listed on Appendix I and Appendix II that is significant and contentious for southern Africa in general, and Zimbabwe in particular. Species can be moved from Appendix I to II under certain circumstances, but so far very few have been down-listed in this way.

The Parks Department in Zimbabwe has argued that the distinction is arbitrary, since Zimbabwe's elephants can in no way be defined as endangered; rather, there are too many of them (DNPWLM, 1991: 5–7). For Zimbabwe, the Appendix I listing is meaningless and is more representative of the opinions of the politically dominant groups in CITES. In turn, pro-ban parties assert that Zimbabwe is defining its elephant population as a national resource, when elephants should be considered in terms of the continental population. In sub-Saharan Africa as a whole, elephants can be defined as endangered, but some areas have an abundance, while other areas were heavily poached and have not yet recovered. Debates over appendix listings reveal divisions over the definition of what constitutes endangered, partly because a species can be defined as endangered on a global, continental, regional or national level.

Nadelman argues that the operation of organisations in international society has homogenised and globalised norms and that, in general, these norms have been influenced by Western ideas and philosophies (Nadelman, 1990: 481). CITES operates two international norms – regulation of trade and prohibition – and the trade in ivory and rhino horn was first regulated and then prohibited. At a single stroke, a trade ban creates an international norm, because it relies on the idea that only halting all trade and consumption will allow a species time to recover its numbers (Princen and Finger, 1994: 131–4). Nadelman argues that parties to a convention adhere to norms out of habit, because it coincides with their own interests or because they fear the consequences of defiance (Nadelman, 1990: 480). The operation of norms has had an effect on the behaviour of parties to the convention. For instance, although a number of Zimbabwean conservationists openly state that Zimbabwe should leave CITES, and that the institution has lost all credibility, Zimbabwe has not yet done so. Any country would find it difficult to leave CITES, because of the political influence of the Northern states and green NGOs. Zimbabwe has so far not deviated from CITES rulings, because to do so would incur the wrath of the international green movement, and this would inevitably result in the removal of conservation aid by NGOs and bilateral aid by governments pressured by the large green NGOs. In addition, Zimbabwe would lose the prestige it has gained and would jeopardise hopes of effectively presenting its case at future CITES conferences.

The operation of global norms through CITES has caused splits in the convention, because the difficulty with global norms is that they do not take account of differing local conditions. International laws, such as CITES, tend to treat wildlife as though it were in a social, political and economic vacuum, but wildlife in Zimbabwe is directly tied to local questions of land distribution and ownership. Zimbabwe has strongly argued for a re-examination of the role of CITES, with Dr Willie Nduku, Director of the Parks Department, stating that CITES had come to behave less like an organisation designed to facilitate and regulate trade, as it was initially intended to, and instead it was acting more as a conservation organisation. The dominance of industrialised states has been resented by many developing states which view it as another institution where their wishes have been overridden by those of the West. Nduku stated that he believed that CITES was dominated by Western industrialised states, and argued that the West had already lost its wildlife and so did not understand the pressures facing developing countries with an abundance of wildlife.[1] In addition, Deputy Director Rowan Martin approached the CITES secretariat in 1993 on the issue of reviewing the convention. However, he had to find a Western state to propose and partly fund the review because Zimbabwe's policies were unpopular in CITES and any suggestions for review which came from them would be immediately viewed with the suspicion that it was intended to unravel the whole CITES system. Finally, Canada agreed to propose a review, which was to be jointly funded by the USA and Canada. The demands for a re-examination of CITES revealed a significant division in CITES, and Rowan Martin felt strongly that Zimbabwe could not propose a review, because they would immediately be shouted down by Western states and green organisations.[2]

An important condition for the creation of a global prohibition regime is the influence of moral proselytism (Nadelman, 1990: 481–3). In the case of CITES, the role of moral entrepreneur has been played by NGOs, such as the International Union for the Conservation of Nature (IUCN) and WWF-International, which were critical actors in lobbying for the creation of CITES. Consequently, NGOs have retained a unique and powerful position in a convention that is essentially an organisation patronised by states. For example, biennial meetings are held to evaluate progress and consider new proposals, and NGOs and trade organisations are allowed to take part in the discussions, but are not allowed to vote on policy decisions (WWF, undated; Princen and Finger, 1994: 135–8). NGOs have been central in arguing for trade bans as a moral issue to save certain species, such as elephants, whales, rhinos and tigers. Clearly, CITES is not simply an organisation that is based around interstate diplomacy, since it involves multiple actors and complex bargaining between different interest groups.

---

[1] Interview Willie Nduku, Director, Department of National Parks and Wildlife Management (DNPWLM), 7.7.95, Harare; see also *Daily Gazette* 14.6.94 'Nduku Criticises CITES'.
[2] Interview with Rowan Martin, Assistant Director: Research, DNPWLM, 29.5.95, Harare.

The divisions between member states have contributed to the perception of CITES as ineffective. The failure of CITES to meet expectations has also been explained by critics as the result of the lack of enforcement capability. International regulation of trade between countries is problematic because it relies on individual nations enforcing the convention effectively, and CITES is poorly enforced across the globe. One WWF report, *Making CITES Work*, found that there was a failure among parties to have adequate national legislation, deficient trade controls and a failure to provide timely and accurate reports on infractions. In addition, it noted that the record of developing states has been no worse than that of wealthy nations, which do not have the excuse of lack of resources (Nash, [1994?]; see also WWF, 1982: 289–94). Many members of CITES fail to provide adequate funding for domestic authorities to ensure proper enforcement. For example, the USA has some of the best legislation to enforce CITES, in the form of the Lacey Act and the US Endangered Species Act. However, the body responsible for implementing that legislation, the US Fish and Wildlife Service, has been starved of funding, due to public sector cost-cutting exercises. The USA is a major importer of illegal wildlife products; however, in 1993, just eighty officers had primary responsibility for 70,000 reported wildlife shipments (EIA, 1994: 6–8). This brings the commitment of CITES member states to ending the illegal wildlife trade into question, since the demand for wildlife products in the world's wealthy states completely overwhelms the existing enforcement structure.

The problems with enforcement arise from the CITES system itself. CITES has suffered from financial difficulties. The budget for CITES as a whole is inadequate and, as a result, WWF suggested that the 1994 budget of 100,000 Swiss francs should be doubled (WWF, 1994b: 5). As early as 1979, Ian Parker carried out an influential study on the ivory trade, which concluded that a system of permits and licences was only as good as the people or institutions that issued them. CITES had created a demand for black-market permits, which could easily be purchased from corrupt officials, and identification of ivory as legal (rather than illegal) was a problem that allowed illegal ivory to enter the legal ivory trading system with forged as well as genuine permits (Parker, 1979: 221–4). It was clear that the mechanism through which CITES claimed to regulate the trade in certain species was flawed in itself, because permits were regularly forged and genuine ones were easily bought from officials.

One of the ways of accommodating the diverse interests of members is to rely on voluntary membership and enforcement. As with all international agreements, parties to CITES are not compelled to belong to the convention. CITES allows all member states to leave the convention or object to any of its provisions. This means that, if a member state objects, it can refuse to abide by the decisions of CITES by taking out a reservation at the time of the decision, but, if a reservation is ever reversed, it cannot be reinstated. Hunting trophies were made special exceptions under

Resolution Conf 2.11 in 1979 (Wijnstekers, 1994: 13). These provisions are considered by some conservationists to be trade loopholes, and the complexities and exceptions to CITES controls make it even more difficult for those responsible for enforcement to carry out their duties properly. They also provide member states with the opportunity to hold CITES to ransom by threatening to leave and trade outside the Convention. This raises questions over the powers of the Convention, since membership is voluntary and can be terminated at any time. Nevertheless, reservations also mean that the particular interests of members of an organisation of over a hundred parties can be accommodated. Despite numerous disputes with other parties, the southern African states have voluntarily remained members of CITES.

Critics of CITES have viewed this dependence on voluntary compliance as a serious weakness, which means that CITES lacks any real powers. For example, Robin Pellew, the director of WWF-UK, stated that the fundamental reason for the failure of CITES was a lack of political commitment. Rather than having real teeth, in the form of an ability to impose sanctions on the offending country, 'a weasel worded statement is patched together that administers a gentle slap on the wrist which the offending countries can brush aside'.[3] Of course, WWF has strong reasons for criticising the failings of CITES and its parties, because it is primarily concerned with conservation of wildlife, while states that are parties to CITES have other considerations. The criticisms levelled by WWF perhaps reflect that the Convention has not made decisions in accordance with their campaigns and philosophies. The USA, Britain and green NGOs, such as WWF-International and the Environmental Investigation Agency (EIA), have argued in favour of better law enforcement to improve the effectiveness of CITES (EIA, 1994).

The rhino horn trade

The rhino horn trade provides an illuminating case study of the politics of CITES and trade bans. CITES is controversially viewed in Zimbabwe as an organisation that primarily seeks to ban trade in wildlife products, rather than one which acts as a regulator, despite more recent indications that CITES may be prepared to consider alternative options to an outright trade ban for ivory and rhino horn. The Parks Department has made clear its desire to reopen a legal and controlled trade in rhino horn. However, the mechanisms of CITES have thus far prevented the Parks Department from doing so. The issue of rhino horn trading has become an important moral theme for parties to CITES and related NGOs. Consequently, deviancy from the provisions of the convention is represented in terms of

---

[3] Robin Pellew, Director of WWF-UK, cited in *WWF News*, Summer 1994, 'Special Report on CITES', p. 10.

moral outrage and evidence of the attraction of financial profit over saving endangered species, and so the norms implicit in CITES have strictly controlled the behaviour of potentially deviant states, such as Zimbabwe.

The continuing illegal rhino horn trade is often cited as the prime example of the failings of CITES and of the negative effects of a ban on legal traders in wildlife products. The trade in rhino horn has been banned since 1977. However, between 1970 and 1990 the world rhino population declined by approximately 80 per cent (Cumming et al., 1990). Meanwhile, the prices for rhino horn have increased dramatically since 1975. What is clear is that the general trend of prices was upwards in the post-ban period: between 1975 and 1980 alone, the price of rhino horn experienced a twenty-one-fold increase (IUCN, 1980: 26). This rise in prices was not accompanied by a shortage of rhino horn, as is usually the case with such staggering increases; in fact, the supply expanded (IUCN, 1980: 27–35). Accurate prices are difficult to obtain, because the trade is illegal and dealers have also been unwilling to reveal how much they paid for rhino horn and how much they have stored away; even when rhino horn trading was legal, dealers under-reported the volume of trade. For example, in the period 1955 to 1980, trade records reveal that the declared exports from Kenya and Tanzania to Japan were between four and twelve times lower than the declared imports into Japan over the same period (Leader-Williams, 1992: 18).

The rise in demand and increase in rhino horn prices was closely linked to illegal trading syndicates, which stretch across Asia and into southern Africa. The involvement of powerful and influential criminal cartels, such as the Mafia, triads and Yakuza, contains problems for law enforcement. Due to the low penalties, lack of enforcement and potentially large profits to be made, drug cartels were also thought to be increasingly involved in the wildlife trade as an attractive alternative (EIA, 1994: 4). Taiwanese dealers have also played a significant role in keeping this trade open. Rhino horn commanded such high prices that dealers viewed buying horn as an investment, and it has been reported that 10 tonnes of horn are stockpiled in China and Taiwan (IUCN, 1992b: 75–6). There is also evidence that dealers were hoping to capitalise on the value of these stockpiles by making the rhino scarcer. When dehorned rhinos were killed in Zimbabwe, one of the possible explanations was that Taiwanese dealers had ordered the poaching cartels in Africa to kill all rhinos. In a sense, they were betting on extinction, because, if the rhino is made extinct by poaching, the value of these stockpiles will rise even further.[4] Recent research by EIA suggests that the rapid price rises for rhino horn in the late 1980s were engineered by a small group of Taiwanese businessmen, who wished to increase the value of their stock-

---

[4] Interview with John Gripper, Sebakwe Black Rhino Trust, 14.10.94, Ascott-Under-Wychwood; *Herald* 17.1.94 'Intensify International Bid to Save Rhino'.

piles, so they sold most of Taiwan's rhino horn to dealers in Hong Kong, who then sold it on to the Chinese dealers. This encouraged smugglers to buy up horn in southern Africa, evidenced by the increased number of Taiwanese citizens arrested in South Africa and a sharp increase in poaching activity after 1988 (EIA et al., 1993: 2–3). So far the dealers have managed to avoid all attempts to halt the illegal trade. Such dealers and criminal cartels are not known for abiding by the law, and any resumed legal trading would have to find a system that effectively makes illegal trading too risky for illegal dealers.

The high value and constant demand for rhino horn has made the rhino a target for poachers and dealers. However, these same two factors, profits and demand, are also what makes rhino horn trading so attractive to southern African range states. The demand from states in East Asia and the Middle East is rooted in deeply held cultural beliefs. The continued demand from these consumers is important, because these states have been targeted as future legal trading partners by southern Africa, and the reasons for demand for rhino horn need to be more fully explored.

East Asia has been the most important consuming region for rhino horn because it has had a long history of demand and cultural attachment to rhino horn for decorative carving and as a medicine. For centuries, the horn has been the most valuable part of the rhino (Bradley Martin and Bradley Martin, 1982: 53–5), but it was the horn's medicinal properties that made it a desirable commodity in the twentieth century. Changes in the world economy after the Second World War meant that East Asians were more and more able to afford traditional Chinese medicines. In China, Mao Zedong's promotion of traditional medicine increased demand for wildlife-based medicines.

Contrary to popular views, rhino horn has never been used as an aphrodisiac. Bradley Martin suggests that this myth began with traders, conservationists and travel writers, who bought rhino horn from the Indian traders they encountered in East Africa. The European visitors wished to know why the Chinese wanted horn and it is thought East African traders stated that it was an aphrodisiac, with an eye to increasing demand for the horn in Europe (Bradley Martin and Bradley Martin, 1982: 66–75). This aphrodisiac myth has since been perpetuated and is often repeated in conservation organisation campaigns to shut down East Asian markets for horn. Rather, it is the medicinal properties that make it so attractive. For a long time the properties of rhino horn were denied by Western medicine, and conservation organisations however, claimed that consuming medicine that contains rhino horn is the same as biting your fingernails. More recently, however, it has been accepted that rhino horn has the same properties as aspirin, which is used to treat common ailments, such as headaches and fevers. In the text on which much of Chinese medicine is based, the *Pen Ts'ao Kang Mu*, its main uses are cited as for reducing fever, for heart complaints and for other preventive purposes (IUCN, 1980). This is similar to Western medicine's use of aspirin to bring down fevers and as a blood thinner for heart patients.

Demand for the horn has arisen wherever there is a Chinese community, but the major markets remain China, Hong Kong and especially Taiwan. As Hong Kong, Macao, Japan and Singapore tightened up their wildlife trading laws during the 1980s, Taiwan took their place. This was only possible because of changes in the world economy, which witnessed the growth of East Asian economies (EIA et al., 1993: 2). The growth in these economies coincided with a rise in demand for rhino horn: more people could afford medicines and rhino horn was bought as an investment. Taiwan has developed a large demand for Chinese medicines, and hence rhino horn (Bradley Martin and Bradley Martin, 1982: 103). Taiwan is a type of black hole for international legislation, because of its special position in the international system, which arises from its history and relationship with China, meaning that Taiwan cannot be named in agreements and resolutions in CITES except as a part of China. However, Taiwan has failed to implement international legislation regarding trade in endangered species. For example, Taiwan holds some of the largest stockpiles of rhino horn, which were accumulated before the ban in 1985, but the authorities refused to destroy them in 1993 (EIA et al., 1993).

The rhino horn trade ban created an international norm that had to be applied at local and national levels. Although the Zimbabwean government was a keen supporter of the ban when it took effect in 1977, its policy position has changed. Graham Child, the director of the Parks Department in 1977, enthusiastically supported the ban at the time. However, he stated that he misjudged the situation and deeply regretted not fighting the ban. Like many southern African conservationists, he argued that sustainable use was the only option to ensure the continued survival of big mammals in Africa.[5] Southern Africa contains the last significant black and white rhino populations. In 1992, Zimbabwe, Namibia and South Africa contained 90 per cent of the continent's black rhino and they have all stated their interest in a legal trade in rhino products (DNPWLM, 1992b: 63).

The theory behind Zimbabwe's rhino conservation policy is based firmly within the framework of sustainable utilisation. The rhino has a potential economic value as a tourist attraction, as a hunting trophy and as a source of rhino horn. In practice, the only value the rhino has is as a tourist attraction, because all consumptive uses of rhino have been disallowed by the CITES Appendix I listing. The Parks Department has been keen to capture the full value of the rhino, and the Department argued that money generated from sport-hunted rhinos and sales of horn would contribute to conservation budgets, which, in turn, would provide for greater protection of rhino in the wild. In terms of environmental economics, consumptive uses of the rhino are presented as a trade off.

---

[5] Interview with Graham Child, Director, SAVE African Endangered Wildlife, and former Director of the DNPWLM, 27.2.95, Harare; *Economist* 9.10.93 'Save the Rhino,' p. 20.

The potential trophy fee for a rhino could be as high as US$250,000, and the auction of one or two rhino hunts per year would contribute significant revenue to failing conservation budgets in Zimbabwe.[6] While the Department has stated its desire to see an end to illegal trade, it has argued that a legal trade conducted through government channels could be used to enhance conservation. The money raised from legal government sales of rhino horn, in theory, would be directly used for rhino conservation in the field (DNPWLM, 1992c). In addition, the potential funds from rhino horn sales could reduce Zimbabwe's dependence on external donors, and so free the Department from pressures to preserve their wildlife, which would then allow the Department more autonomy in decision-making (DNPWLM, 1992b: 63).

Controversially, Zimbabwe does not want the markets for rhino horn in the Far East to be closed. Although these markets are illegal and have so far contributed to the sharp decline in Zimbabwe's black rhinos, to close them down would effectively deny Zimbabwe any future legal trading partners. In fact, Zimbabwe has been accused of blocking CITES proposals to initiate trade sanctions against states that consume rhino horn. The Southern African Group for the Environment (SAGE) and EIA have stated that Zimbabwe and South Africa joined forces to prevent sanctions being imposed on Taiwan and China at a critical meeting in 1994 at Geneva on tigers and rhinos.[7] However, eventually, in 1994, trade sanctions were applied against Taiwan by the USA, valued at US$25 million.[8] This was a largely symbolic measure, as the amount involved did not have a serious economic effect on Taiwan.

Another explanation for Zimbabwe's policy position is that the rhino horn trade ban has manifestly failed to halt the decline of the rhino, and in fact the Parks Department claimed that the ban was responsible for an increase in poaching in Zimbabwe.[9] The Parks Department argues that the ban has failed, so it is time to try another strategy, and that this could involve some consumptive and non-consumptive utilisation of rhinos. In fact, traditionally deep-green and preservation-orientated NGOs have begun to suggest that the failure of the ban could force a change in international policy on rhino horn trading (WWF, 1992: 2–3; Milliken et al., 1993: 55). The international ban on the rhino horn trade is a highly contentious issue, and it is clear that it is not purely a matter of applying neutral conservation principles. Zimbabwe's policy philosophy deviates from the norms established by CITES, and the idea of utilising rhinos as hunting trophies, farming them for horn and legally trading horn has been met with vociferous opposition. Conservation NGOs, acting as moral entrepreneurs within CITES, have actively opposed the

---

[6] *WWF News* Winter 1992–3 'On the Horns of a Dilemma'; DNPWLM (1992c: 19).
[7] *Financial Gazette* 7.4.94 'Zimbabwe Slated Over Rhinos'.
[8] *Guardian* 21.10.95 'Horns of Plenty'.
[9] *Herald* 13.1.93 'Sixty Four Held For Rhino Horn Deal'.

Zimbabwean Parks Department and conducted international publicity campaigns against them. Fear of the consequences of deviance from the norms in CITES have prevented Zimbabwe from converting its conservation philosophy into policy with regard to the rhino.

The ivory trade

The case of the ivory trade is related to that of the rhino horn trade. As with rhino horn, Zimbabwe's policy position is to seek to resume trading, but, unlike the rhino horn ban, Zimbabwe did not willingly accept the ban, but lobbied hard against it. The ivory trade differs in a number of respects from the rhino horn trade. Firstly, there is no disagreement over the status of the rhino, and the fact that it is an endangered species is widely accepted. Yet there has been substantial disagreement over what constitutes endangered with regard to the African elephant. Again, this is related to major differences between opposing environmental ideologies, where critics of sustainable utilisation have argued for adherence to the precautionary principle, while Zimbabwe and its supporters have promoted sustainable utilisation for conservation. The second difference is that, while there is a degree of cultural attachment to ivory in East Asia, demand for rhino horn is based on a cultural demand for medicines. In contrast, ivory is required for luxury items, such as jewellery and ornaments, it is easily substituted by other materials, and it is subject to changing fashions, so, in principle, the ivory trade should be more easily controlled or shut down because fashions can be changed. For southern Africa, the ivory trade is a key policy issue, and the ban highlights important divisions between the interests of an international organisation and particular nation-states. It demonstrates the difficulties that an international system faces when it attempts to regulate the use of wildlife held within national borders.

CITES has evolved into an international prohibition regime because the environmental movement has favoured the precautionary principle and pro-banners have been the most dominant parties. Industrialised states have had the most influential position in CITES, and so the convention reflects Western values concerning legitimate use of animals. Through the CITES ban on ivory trading, the organisation has globalised Western values relating to wildlife conservation and to what constitutes legitimate use of animals (Nadelman, 1990: 484). The NGOs that promote moral and ethical arguments regarding legitimate use of animals have assisted the industrialised states in maintaining their powerful position. In the 1980s animal rights and preservationist NGOs and conservationists suggested that the only way to control the ivory trade was to cut demand. Conservationists argued in favour of a ban on ivory as the only way to stop ivory poaching and the eventual extinction of the African elephant (Care for the Wild [1992?]). The environmental movement argued in favour of the precautionary principle – that to ban the trade was better than to risk the potential extinction of the species. Closing all markets for ivory was presented as an easier option for enforcement than attempting to shut down illegal activities in numerous trading entrepôts and range states. The elephant had become the victim of Barbier's concept of the economics of extinction (Barbier et al., 1990). In effect, poachers, dealers and buyers in the end use markets were gaining all the profits from ivory while the range states did not receive these benefits, because they were unable to establish their control over the resource, due to a lack of investment in wildlife protection and enforcement (Ivory Trade Review Group, 1989; 31–4; Barbier et al., 1990). The previous attempts to regulate the trade had failed to stem the rates of poaching, and so it was recommended that the African elephant be transferred to CITES Appendix I in 1989.

Conservation pressure groups, Western states and East African states formed a powerful alliance in favour of an ivory ban. Pro-ban African states actively courted an alliance with NGOs and Western states in order to obtain conservation funding. In turn, external support for anti-poaching assisted East African governments in another conservation agenda, which was to coerce local people into moving away from prime wildlife tourism areas (Peluso, 1993: 200–9). Tanzania and Kenya were two key East African elephant range states that publicly called for a ban on ivory trading, partly because they were the worst hit by poachers and argued that only a total ban would save their elephants. Richard Leakey of the Kenya Wildlife Services was especially prominent in calling for a ban, since Kenya had been hit hard by poaching and this had tarnished its reputation for wildlife conservation and its attractiveness to Western tourists.[10] It was clear that Kenya had to improve its image and assure Western conservation donors and organisations that it was not in favour of the trade, which was destroying its tourist industry and conservation record. In order to improve its image, Kenya publicly burnt 12 tons of ivory, worth US$3 million, while rumours circulated that Leakey had been offered a large sum from an anonymous donor in return for burning the ivory (Bonner, 1993: 151). The publicity stunt worked and, as intended, conservation aid poured in (Milliken, 1994a). Leakey's extensive restructuring of the corrupt Kenya Wildlife Services impressed the World Bank and other donors so much that they approved grants and loans for conservation. Consequently, Kenya was in the fortunate position to be able to develop a donor-dependent conservation policy based on preservation.[11] The aid offered to Kenya for its pro-ban stance demonstrated the position that was taken by the majority of Western states and conservation organisations. In contrast, southern Africa was the proponent of continued, properly regulated ivory trade and was

---

[10] *Guardian* 20.7.95 'The Man Who Would Be Kenya'; Woods (1999: 31–3).
[11] Interview with Raoul du Toit, DNPWLM/WWF, 7.5.96, Harare.

vilified for its anti-ban policies. Pro-banners were hailed as elephant saviours, who argued that a continued legitimate ivory trade would only provide a screen behind which the illegal trade could hide.

Opposition to the ban on ivory has stemmed from a fundamental disagreement over the best means of saving elephants and over the precise meaning of the term endangered species. Opponents of an ivory ban formed an alliance within CITES to lobby in favour of their policy position. The southern African pro-trade states have received support from Canada and Norway, whose sealing and whaling policies are constrained by international agreements in the same way as African elephant policies, and similarly, as the world's largest single ivory consumer, Japan has continued to be allied with the southern African states on the ivory trade issue.

Nadelman suggests that international prohibition regimes are more likely than other regimes to ignore the considerations of costs and consequences of decision-making and that, like crusades, they are pursued even when alternative approaches appear less costly and more effective. Deviance is rarely tolerated by international prohibition regimes, because it has the potential to undermine the regime (Nadelman, 1990: 524–5). Since the establishment of CITES, the alliance of preservationist states has wielded the most influence. This has meant that Zimbabwe and its allies have had difficulty in presenting sustainable use and controlled trade as effective methods of conservation and, as a result, pro-trade parties have been relatively isolated in CITES. The problem for pro-ban parties is that Zimbabwe holds one of Africa's largest elephant populations and is one of the last strongholds of the black rhino. The problem for Zimbabwe is that it is still reliant on conservation aid and other economic aid, and these resources are held by pro-ban parties and NGOs that favour preservation. As a result, the norms implicit in the CITES ban – that killing elephants for ivory will result in serious depletion of the species have had a profound and direct effect on the policy choices available to Zimbabwean wildlife conservation agencies.

Zimbabwe's policy position is to lobby for a resumption in the legal ivory trade arguing that legal trading would generate income for conservation budgets to support elephant conservation. Rowan Martin claimed that he was being approached by Asian dealers interested in the ivory held in the central store and that generally offers were around US$1,000 per kilo, which meant that the 30-ton stockpile was worth US$30 million (in 1995).[12] As ivory commands such a high price in hard currency, it could provide large amounts of revenue for conservation. In this sense, ivory represents Zimbabwe's comparative advantage in the international trading system, and yet Zimbabwe is blocked by international legislation, which means it cannot derive the total potential economic benefit from elephants (Barbier et al., 1990). Consequently, Zimbabwe's inability to

---

[12] Interview with Rowan Martin, Assistant Director: Research, DNPWLM, 29.5.95, Harare.

sell ivory and elephant products has been presented to the international community as an unacceptable waste of a national resource. Willie Nduku commented that he felt it was wasteful to destroy something that is continually produced by natural death, by 'problem animal control' operations and by culling, especially when that product is required by other states, such as Japan.[13] Zimbabwe has come to be viewed as the new hard-liner on this issue, since it refuses to modify its position.

One of Zimbabwe's key anti-ban arguments is that such a sweeping international policy ultimately hurts poor rural communities, especially by denying rights to utilise wildlife under the Communal Areas Management Programme for Indigenous Resources (Campfire). The international community has found this argument the most difficult to resist. Taylor and Cumming argue that elephant management policies must be politically possible, socially acceptable, technically feasible, ecologically sustainable and economically viable, and the ivory ban, they suggest, does not fulfil these criteria (Taylor and Cumming, 1993). At the time of the ban, Zimbabwe was embarking on its Campfire policy, which revolves around the idea that wildlife should be sustainably used for the financial benefit of the community. Rowan Martin argued that the ivory ban would result in a loss of revenue for Campfire areas, since they would be disallowed by the international community, from selling their most lucrative resource, and that, if wildlife was not an economically viable form of land use for communal farmers, they would simply swap it for another land use that was viable.[14] The debate surrounding the effectiveness of Campfire as a rural development and wildlife conservation strategy has proved to be the most powerful political weapon in Zimbabwe's argument in favour of a renewed trade in ivory. Opposers are characterised as misanthropes who seek to deny the rural poor a chance of earning a living from animals that are otherwise, at best, a nuisance and, at worst, crop destroyers and a threat to human life.

Lal argues that, in the post-Cold War world, globalisation of Western environmental ideas through green NGOs has constituted a new form of imperialism. The difficulty is that ecofundamentalism is presented in terms of rational science, while they are at heart moral crusades (Lal, 1995). It is clear that conservationists in Zimbabwe view the international green movement as having played an unwelcome and restrictive role in their policy formulation. The experience of Kenya during the height of poaching demonstrated the devastating effect that adverse publicity could have on the economically important tourist industry. International NGOs have the links to international media that could result in a damaging campaign to boycott Zimbabwe as a tourist destination, because of its conservation philosophy. One of the main arguments for the ban was that

---

[13] Interview with Willie Nduku, Director, DNPWLM, 7.7.95, Harare; *Environment Bulletin* Dec. 1994–Jan. 1995 'Battle Commences Over the Ivory Trade'.
[14] *Herald* 16.12.94 'Allow Trade in Legally Obtained Ivory and Rhino Horns: Campfire'.

it would stop poaching by removing the market for poached ivory. However, the Parks Department has always disputed this reasoning, arguing that bans only drive up the price of ivory and rhino horn. A number of conservationists in Zimbabwe stated that the ivory ban was directly responsible for an increase in elephant poaching. The feeling that the international community misunderstood the southern African position was particularly acute because Zimbabwe had a relatively good record on poaching in the 1990s, and it was felt that southern Africa was being punished for East African elephant management failures. This sentiment translated into charges of neocolonial behaviour against pro-ban parties at CITES and that the ban was an unjust international action.[15]

The notion that Zimbabwe has suffered from a lack of recognition for its success in wildlife conservation is deeply felt, because Zimbabwe is regularly denounced for elephant (and rhino) management failures, such as culling and sport hunting, by the same Western conservation organisations that congratulate the success and protectionist policies of East African states, such as Kenya. Indeed, Brian Child commented that Zimbabwe administered and managed its ivory stocks properly, but received no recognition for its good record.[16] In response to the need to strengthen the bargaining position of southern Africa in CITES and to a continued trade ban in 1991, Zimbabwe, Malawi, Botswana and Namibia joined together in a pro-trade alliance, the Southern African Centre for Ivory Marketing (SACIM), now renamed the Southern African Convention for Wildlife Management (SACWM) to reflect its broader role in conservation. The SACWM agreement was significant for CITES, because SACWM states contain approximately 150,000–200,000 of Africa's elephants (WWF, 1994). The SACWM treaty has its origins in the polarisation of the debate over the ivory trade, because the SACWM states insist that the ban on the ivory trade frustrates long-term and effective elephant management in southern Africa. SACWM has a common policy, involving sustainable utilisation of elephants (including a regulated legal trade in ivory) and community involvement in conservation (on the Campfire model) (DNPWLM, 1992a: 10; SACIM, 1994). So far, SACWM has had little power to change the international community's position on ivory trading, but it has continued to argue the southern African case. South Africa has also expressed an interest in joining SACWM (as it is not covered by the CITES ban) and, if South Africa joined SACWM, it would be strengthened and able to declare a reopening of the ivory trade. The problem that SACWM faces is that, of

[15] Interview with Keith Madders, Director, Zimbabwe Trust (UK), 12.10.94, Epsom, Surrey; interview with Brian Child, head of the Campfire Coordination Unit, DNPWLM, 16.5.95, Harare; interview with Charl Grobbelaar, Chief Executive of the Zimbabwe Hunters Association, 12.2.95, Harare; see also Dublin et al. (1995: 17).
[16] Interview with Brian Child, head of the Campfire Co-ordination Unit, DNPWLM, 16.5.95, Harare.

the ivory consuming nations, only North Korea has not joined CITES. Nevertheless, SACWM has threatened to leave CITES and trade ivory legally.[17]

The arguments over the ban revealed that the opposition and the protagonists were arguing from differing ideological standpoints. Animal welfare organisations and some conservation NGOs used animal rights arguments to demonstrate that the elephant had a value beyond commercial use. Elephants were presented as sensitive and intelligent creatures that deserved greater compassion and consideration. Preservationists simply suggested that all elephants should be preserved. Zimbabwe, on the other hand, was arguing from a utilitarian standpoint. Zimbabwe claimed that the African elephant could not be considered an endangered species according to CITES criteria for an Appendix I listing. It was argued that Zimbabwe's record on sustainable use of wildlife products had been a great success thus far (even increasing the number of elephants) and so it should be allowed to continue without external interference.

The particular nature of the ivory trade has been the subject of debate at CITES. The way in which legal and illegal ivory sales have mixed in the past has formed the centre-piece of the argument for the continued ban on ivory trading. Pro-banners have argued that a legal trade merely provides a cloak for illegal trading to continue, while pro-traders have favoured tighter controls to prevent corruption.

The African elephant was first placed on Appendix II in 1977, but, from then until the ban in 1989, there were numerous attempts to regulate the trade. It was during the 1970s and 1980s that the demand for ivory became so great that it began to threaten the survival of the species. As a result of the lucrative ivory trade, the numbers of African elephants were halved, from 1,300,000 to 600,000, in the 1980s (Dublin, 1994). During the 1980s, demand for ivory came from the luxury goods market in Europe and the USA, and the longer-standing trade with the Far East. Increased demand also came from the growing economies of the Far East, notably Hong Kong, Japan and China. The legal ivory trade was an important source of income for some states. For instance, it is estimated that, during the 1980s, the ivory trade was worth US$50–60 million per annum to the continent as a whole (Barbier et al., 1990: 36). This was not a large sum when taken in the context of sub-Saharan exports as a whole, but, for a few countries, notably Zimbabwe, South Africa and Botswana, it was a significant source of hard currency. In Zimbabwe, the amount of money retained by Zimbabwe from the ivory trade was high, as, blessed with little poaching, Zimbabwe was able to sell much of its ivory legally on the international market, and such sales generated an estimated US$11 million per annum. In addition, Zimbabwe was home to a thriving

---

[17] *Economist* 5.11.94 'Whose Elephants Are They Anyway?' p. 70.

carving industry, which consumed 30 tonnes of ivory per year and employed 400 people (Cumming and Bond, 1991: 24).

The rise in demand for ivory in the 1970s and 1980s came from Western Europe, the USA and the Far East, but it was the demand from Asia that became extremely important in the pre-ban period, since Japan and Hong Kong accounted for 75 per cent of world imports of ivory (Barbier et al., 1990: 53). Japan was the world's biggest single ivory consumer, as postwar incomes rose. It was reported that Japan alone was importing 1,900 tonnes of ivory per year prior to the ivory ban and that, by 1992, it had stockpiled 100 tonnes in anticipation of a renewed legal trade (EIA, 1992). In Japan, the demand for ivory tended to be for jewellery, for parts of traditional instruments and for small blocks of ivory to be carved into *hanko* (personal seals). Japanese investors also saw ivory as a stable commodity to invest in during the economic upsets of the 1970s and 1980s. They were attracted to it as a store of wealth, because in Japan ivory has cultural value and so its level of demand and price remains relatively stable, whereas in Europe and the USA it is a luxury item, which is subject to peaks and troughs in demand and related price fluctuations (Barbier et al., 1990; Bonner, 1993). It was important to investors to protect the value of ivory in domestic and international markets, and so in 1984 the Japan Ivory Importers Association was established, as importers became aware of the potentially adverse effects of the publicity surrounding the ivory trade. The association has a central fund for conservation and has made donations to CITES in the past (Thornton and Currey, 1991; EIA, 1994).

Hong Kong has also occupied a unique place in the international ivory trade. It was not a major domestic consumer, but instead it was a carving centre, which re-exported ivory to China, Japan, South Korea and Taiwan. Hong Kong became a major entrepôt and ivory stockpiler in the 1980s, because of a legal loophole in CITES, which allowed worked ivory to enter freely, and so from Hong Kong semi-worked or carved ivory could be legally exported to anywhere in the world without permits. In addition, at the time of the ban, Hong Kong was exempt for six months, due to its status as a colony, and Britain and China had taken out a special reservation on behalf of the Hong Kong traders, which allowed them to continue to sell as much ivory as possible. An EIA report claimed that 200 tonnes of ivory went missing (in effect, it was traded) from official stockpiles during this period and, as a result, the official figure of a 670 tonne stockpile has to be revised to 474 tonnes (EIA, 1992: 52).

The ability of the illegal trade in ivory to avoid all controls formed the centre-piece of the pro-ban argument. During the 1980s it was clear that poaching in East Africa was continuing at unsustainable rates, with the most notorious case being Kenya, where Richard Leakey, of the Kenya Wildlife Service, estimated that between 3,000 and 5,000 elephants were lost per year to poachers in the 1980s.[18] One of the major problems with the pre-ban ivory trade was that the actors involved were dispersed across a range of countries and continents. This meant that CITES was unable to cover and regulate all aspects of the trade effectively. One of the most important controls was the ivory quota system, established in 1985. Each state determined their own export quota, related to the size of elephant population and the quality of management, and all existing stocks of ivory were legalised. This controversially legalised two ivory stocks held by Burundi and Singapore (neither of which has an elephant population of its own). CITES was severely criticised for the action (Thornton and Currey, 1991). Indeed, the amount of ivory legally traded in the mid- to late-1980s did decline. From 1979 to 1985, the minimum volume of raw ivory exported from Africa was estimated at 868 tonnes annually, and this fell to 587 tonnes in 1986 and just 176 tonnes in 1989 (Milliken, 1994a). Supporters argued that this was due to the success of the quota system and it just needed more time to work. However, it was more likely that it was the decrease in the average tusk size which accounted for the fall in tonnage traded, and so more elephants were being killed for less ivory (Barbier et al., 1990). The system was criticised as favouring poachers and corrupt officials by allowing them to benefit from sales of illegal stocks on the legal market. There was no obligation for states to submit their quotas in the CITES-recommended format, and very few states that submitted a non-zero quota in 1987 gave them in full format (Cumming et al., 1990). The quota system, when used properly, was nothing more than a statement of intent by producer countries. The CITES control systems were fairly easily avoided by traders, the primary legal loophole in the system being the failure of the quota system to cover worked ivory, since it did not require permits so raw ivory could be carved and was, in the process, instantly made legal (Ivory Trade Review Group, 1989).

The legal trade was conducted alongside a more extensive illegal ivory trade in the 1970s and 1980s. The illegal trade in ivory caused greatest concern and threatened the survival of the species. The extent of the illegal ivory trade is impossible to quantify. The illegal trade was dependent on well-organised and long-established smuggling routes, which stretched from Africa to Europe, the Middle East and the Far East. For example, in 1986, 17,000 tusks, worth an estimated US$3.75 million, were auctioned in Mogadishu, and the sale came after the biggest seizure of illegal ivory in Antwerp. The tusks came from Somalia and were eventually destined for Dubai.[19] This demonstrates the extent of the routes used by smugglers, with Belgium acting as a popular entrepôt in the 1980s for illegal ivory, because of lax customs controls. In Dubai, they would have been carved to exploit the legal loophole in CITES, which

---

[18] Richard Leakey in his address to the Wildlife Society of Zimbabwe AGM 1994.
[19] *Herald* 2.3.86 'Auction Might Save Africa's Elephants'.

defined carved or semi-worked ivory as legal. One of the most striking features of the illegal trade in ivory is its fluidity, as it managed to avoid most regulations, and if one entrepôt is shut down, another quickly opens. For example, in 1984, when Macao began to enforce CITES rules, Singapore (a non-CITES state) took over its role, so that, by 1986, Singapore accounted for 55 per cent of world consumption of raw ivory (Barbier et al., 1990: 68–71). The involvement of triads and the Mafia in the ivory trade has also made it difficult to regulate. In debates about reopening the ivory trade, this factor is often ignored, but the interest that ivory attracts from criminal elements is a significant issue for law enforcement (Currey and Moore, 1994: 3; EIA, 1994: 4).

The involvement of officials in the illegal trade made it especially difficult to enforce CITES regulations. Such involvement neutralised the enforcement capacity of wildlife departments and of international attempts to stamp out the trade. Diplomatic immunity is widely used to protect movements of rhino horn and ivory. The use of privileges associated with the diplomatic bag means that wildlife products can be exported by diplomats with relative impunity. In a rare case when a diplomat was caught, the Indonesian Ambassador to Tanzania was stopped and found with ivory in his baggage in 1989 on his way to Singapore. Likewise, missionaries stationed in Tanzania, who were regarded as law-abiding and honest, were also implicated in smuggling ivory (Thornton and Currey, 1991: 132–47). The involvement of diplomats continued in the post ban period. For example, in 1995, Kenya requested that the United Nations (UN) lift diplomatic immunity on two UN officials (an American and a Sri Lankan) involved in the United Nations Operation in Somalia (UNOSOM), who were accused of smuggling ivory from Somalia to Kenya.[20] The extent of the illegal ivory trade and the poaching it fuelled were real causes for concern for conservationists in the late 1980s. It was clear that the numerous attempts to limit the trade had failed, and so for the majority of conservationists it seemed that a total trade ban was the only solution.

The politics of the ivory ban

There were a number of actors involved in getting the ivory ban on the international agenda, but they formed two broadly opposing sides. Those against the ban were the southern African states and ivory consuming states, which generally promoted the policy of sustainable use as the most effective method of conservation. Those in favour of the ban included Western states, East African states and Western conservation organisations, which generally argued from preservationist standpoints.

---

[20] *Herald* 28.3.95 'Kenya Asks UN to Lift Smuggling Suspects' Immunity'.

The green NGOs argued from a position that was heavily influenced by deep-green and animal rights philosophies. The influence of NGOs over CITES parties came from their central role in establishing CITES and their rights to attend the biennial conferences. In turn, the industrialised states and East African states adopted the language of these organisations in response to pressure from campaigning organisations and their publics at home who wanted to see their government involved in 'saving the elephants'.

Until 1988, the idea of an ivory ban did not hold much currency, since CITES remained convinced that a controlled trade was sufficient to ensure that trade did not threaten the survival of the species (see Parker, 1979: 207). International concern centred around whether the elephant was in danger of extinction from the trade. Zimbabwe argued that the elephant could not be considered as endangered, while the opposition argued that to fail to institute a ban might lead to extinction. Multiple actors were involved in the decision to ban the ivory trade in 1989, and the methods used by the pro- and anti-ban interest groups in part explain the depth of feeling and high profile clashes that surrounded debates over the ban. From the point of view of anti-ban interest groups, those lobbying in favour of a ban were more concerned to placate domestic constituencies in industrialised states than with the need to examine the complexities of large mammal conservation in developing states.

International NGOs were critical actors in the decision to ban the ivory trade. They filled a diplomatic niche left at the end of the Cold War for non-state actors. Diplomacy related to urgent human rights and environmental issues was left to NGOs, rather than to nation-states, because traditional models of diplomacy have proved unable to respond to urgent threats that require rapid agreements and solutions (Princen and Finger, 1994: 29–32). Consequently, NGOs have occupied that space. The late 1980s and early 1990s were the heyday of the international environmental movement, and coincided with renewed debates over an international ivory ban. Princen and Finger argue that the case of the ivory ban constituted a form of resistance to dominant economic forces (Princen and Finger, 1994: 121–59). NGOs were moral actors that argued from ethical standpoints for an ivory ban, and their ability to communicate moral arguments to the public in industrial states ensured that ivory became a socially unacceptable commodity. NGOs were not constrained by any particular national interests, and were able to present themselves as operating from a moral standpoint drawn from genuine concern for elephant welfare. Elephants and rhinos were high profile flagship species for the conservation movement partly because they are large mammals that are easily identified by conservationists and their supporters in the West. Elephants, in particular, can be easily endowed with human-like characteristics, such as caring, grieving, sensitivity and being family-orientated. The popularity of this type of representation of animals in films and cartoons is evident in *Dumbo*, the *Lion King*, *Free Willy*, *A Bug's*

Life and *Babe*, and our screens are filled with such images of talking and feeling animals from early childhood.[21] Such anthropomorphic representations of elephants have made their continued survival an area of heated debate. The public in industrialised states have become accustomed to viewing certain mammals, such as whales and elephants, as having comparable intelligence and consciousness to those human beings. As a result of such attachment to these animals, their survival has become an industry that brings in funds to conservation organisations and states alike. The continued wrangles over how best to protect them carry multiple interests with them, and arguments from the varied interest groups resurface at the biennial conferences of CITES.

The role of NGOs in the ivory ban was so significant that Zimbabwe and other pro-trading interest groups have accused them of environmental imperialism. Eugene Lapointe commented that the 1989 meeting of CITES witnessed the explosion of the animal rights movement on to the international scene. He claimed that it dominated decision-making at that meeting and that the ten or twelve groups involved turned a forum for discussion into a farce, deprived states of their decision-making power and began the era of international blackmail in conservation.[22] It is very clear that NGOs had an important role in manipulating the media and public opinion to change the pro-trade stance of certain states and organisations. For example, EIA produced a highly influential report, entitled *A System of Extinction: The African Elephant Disaster*, which was circulated to all CITES parties. It denounced the pro-trade position, accused CITES of presiding over the extinction of the elephant and argued that Zimbabwe's policy position was determined by the need to protect domestic interests in the culling, hunting and ivory industries (EIA, 1989). Their role as public opinion manipulators was especially significant, because NGOs were not allowed to vote in CITES meetings, and so, while they had no votes, they did have a definite input into the decision-making process. In addition, if organisations could not change policies by persuading their home governments to pressure range states, NGOs could use their own conservation aid to induce states to follow the policies they perceived as best. Rowan Martin was concerned that conservation organisations have been the worst for offering aid with complex strings attached, and WWF-International and IUCN have both used conservation aid to push for their policies. In addition, it was reported that Rowan Martin was approached by the US delegation to CITES during the 1989 meeting and reminded that he should think of the aid that Zimbabwe received from the USA and World Bank when voting on the ivory issue (Bonner, 1993). Indeed, one conservationist in

---

[21] *Independent* 26.2.96 'When An Elephant Gets Emotional'.
[22] Eugene Lapointe, Secretary General of CITES, speaking at the Symposium on the Conservation and Sustainable Use of Natural Resources, held in Tokyo, 1 October 1993, Global Guardian Trust.

Zimbabwe commented that the American delegation at CITES could be 'really mean' if they did not get what they wanted.

The African Wildlife Foundation (AWF) can claim to have had a critical role. As late as 1988, there were no calls for an ivory ban. However, in 1988, AWF sent out an appeal to raise money and declared 1988 the Year of the Elephant. In their press conference they urged the American public not to buy ivory, but they still did not call for a total ban. The press conference aimed to shock the public about ivory poaching and jolt people into stopping buying ivory. The publicity, which talked in terms of a crisis facing elephants, made effective use of television and other media. Indeed, Zimbabwean Government documents noted the influence of the campaign on attitudes to ivory and Zimbabwe's policy (Martin et al., 1989). The campaign also represented an opportunity to increase funding for AWF elephant conservation programmes. The publicity caught the attention of a number of animal rights organisations. According to Bonner, the strongest attacks on those who opposed the ban came from the US-based Friends of Animals, which urged people to write to WWF denouncing their support for continued ivory trading (Bonner, 1993). By 1989, WWF caved in, fearing the loss of a large section of its membership, and thereby its funding, because by then elephants had become *the* environmental fad.

EIA also claims to have played a critical role in the campaign for an ivory ban, and certainly, from Zimbabwe's viewpoint, this is the case. Its investigation into the international illegal trade in ivory was financed by Christine Stevens of the US-based Animal Welfare Institute. The information it obtained on how illegal ivory was laundered into the legal trading system and on the extent and organisation of poaching was crucial in opening the world's eyes to the problem. EIA was a skilful media manipulator. For example, it filmed Daphne Sheldrick's famous elephant and rhino orphanage in Tsavo National Park. It was convinced that, if the British and American public saw these 'tiny casualties' of the poaching war their hearts would be won over (Pearce, 1991: 68–72; Thornton and Currey, 1991: 82). The footage on poaching in Kenya was used by Independent Television News (ITN); these powerful video images snowballed, along with other media coverage of the ivory issue, and afterwards the media attention focused the public mind on the ivory trade and ban. However, in some ways, WWF took the credit for bringing the situation to light, and it was inundated with letters of support and funds (Thornton and Currey, 1991: 200–14). Organisations used shock tactics to convey their message, and denounced those who opposed them as unconcerned with the plight of the elephant. This was not unusual amongst campaigning organisations, which recognise that creating an air of urgency through shock tactics is the only way to get their particular issue into the public mind and on to the political agenda.

EIA also teamed up with WWF-USA to draw up a proposal to ban the ivory trade. However, EIA was aware that the proposal would be more credible if it was actually proposed by an African range state. As Kenya's conservation record was too tarnished, Tanzania was persuaded to propose it (Pearce, 1991: 73; Thornton and Currey, 1991: 195). EIA set about collecting information to prove that Zimbabwe's claims of an abundance of elephants were false and used the fact that Zimbabwe and apartheid South Africa had a unified position to criticise Zimbabwe. EIA accused Zimbabwe of double-counting – counting elephants before they crossed to Zambia and Botswana and counting them again when they returned. In addition, it claimed that Zimbabwe's assertion that its elephant population was increasing at 5–7 per cent per year was biologically impossible, and that such an increase was due to migration to the safety of Zimbabwe from heavily poached areas in Zambia. Lastly, EIA argued that Zimbabwe's complaints about extensive damage to vegetation by elephants were exaggerated and inaccurate (Currey and Moore, 1994: 11). Of course, these accusations were rejected by Rowan Martin of the Parks Department, who was representing Zimbabwe at CITES in 1989.

The importance of the ivory ban to NGO funds was illustrated by the behaviour of WWF. In debates over the ban, international NGOs realised that, in the long term, the fate of their organisation could be determined by its stance on the ivory issue. WWF-International, as one of the biggest conservation organisations in the world, received a great deal of credit for the ivory ban. However, its role was more ambiguous, since national chapters clearly disagreed with WWF-International, because it did not call for an ivory ban immediately. John Hanks, head of the WWF Africa Programme, was closely associated with the position of WWF-International, which remained anti-ban until 1989. Hanks was clearly in favour of sustainable use. The furthest the organisation would go was to support a split listing, meaning that southern Africa's elephants would go on Appendix II and the rest on Appendix I (Thornton and Currey, 1991: 254). Bonner has suggested that the position of WWF-International was determined by the large number of South Africans on its Executive Board and South African funders in the 1001 Club (Bonner, 1993: 69). However, WWF-USA and WWF-UK were more interested in a ban because of their home constituencies. Rowan Martin commented that Zimbabwe's biggest problems come from WWF-USA because their membership is comprised of 'people who want to save wildlife so they cannot support sustainable use'.[23] NGOs supported a ban partly to justify their own continuance, as for WWF-USA and WWF-UK to fail to call for a ban would have meant that memberships, and thus a major block of funds, would have been removed by supporters. Hence, when other animal rights-oriented organisations threatened to tell WWF members that the organisation was supporting the killing of elephants, it started to call for a ban (Pearce, 1991: 73; Thornton and Currey, 1991: 254). WWF, as a whole, is certainly

[23] Interview with Rowan Martin, Assistant Director: Research, DNPWLM, 29.5.95, Harare.

frightened of losing its fund-raising base, and the African elephant attracts a large amount of funding, and so it would have meant serious problems for the organisation if it ignored the sentiments of the membership base.[24] However, in listening to its membership base, WWF ignored the opinions of many elephant range states. Zimbabwe felt betrayed and has not forgotten the role of WWF, which many Zimbabwean conservationists consider to be a pathetically weak organisation in face of opposition from its membership.[25] The disagreements over the ivory ban between the WWF-International and national chapters revealed deep divisions in the organisation over the best means to manage elephants.

Finally, the ivory ban was instituted in 1989. A study by a consortium of NGOs had been commissioned by CITES on the status of the elephant, and the Ivory Trade Review Group (ITRG) report was used effectively by pro-ban parties to support their case. Overall, the ITRG report concluded that a ban was the most effective way of saving the African elephant, and it came at a time when the campaign for a ban was at its peak. However, the report did conclude that elephants in southern Africa were in no way endangered by the trade (Ivory Trade Review Group, 1989: 1–4). Yet the report was promoted as central to the pro-ban argument. Bonner suggests that the timing of the release of the report was politically motivated, because the study was not meant to be released for a further three months, in October 1989 (Bonner, 1993: 142). As a result, the USA, the UK, the European Community (EC), Switzerland, Australia, Canada and the UK on behalf of Hong Kong all announced full or partial ivory bans within two months (Pearce, 1991: 74; Wildlife Conservation International, 1992: 6). These unilateral bans immediately cut demand for ivory and had a devastating effect on the illegal and legal trade in Europe and the USA which had been heavily influenced by NGO campaigns, which meant a failure to vote for a ban would have been politically unpopular at home. The unilateral ivory bans established by the USA and Britain, and their pressuring for a total ban, meant that they could be seen to be doing something about the plight of the elephants. The debate at CITES centred on three proposals: a total Appendix I listing (a ban), a split listing for East and southern Africa, which allowed trade to continue for southern Africa, or a continuance of the status quo. It was clear that some kind of compromise was necessary to prevent southern African states simply leaving CITES altogether. The compromise was that countries could take out a reservation on the ban and apply for a downgrade to Appendix II for their elephants if their management policies met certain criteria. The outcome was seventy six in favour, eleven against and four abstentions, and formal trading ceased on 1 January 1990 (Bonner, 1993: 159). Shortly afterwards, the carving centres in Hong Kong and China virtually shut

---

[24] Interview with Keith Madders, Director Zimbabwe Trust (UK), 12.10.94, Epsom, Surrey.
[25] Interview with Rowan Martin, Assistant Director: Research, DNPWLM, 29.5.95, Harare.

down (Milliken, 1994a: 49–57). Zimbabwe, South Africa, Botswana and Namibia continued to hold reservations and were allowed to trade ivory amongst themselves and with non-CITES parties.

The ivory ban resulted in two years of mistrust and acrimony between southern African states and NGOs. It created deep divisions over conservation practice between East and southern Africa. Ban advocates presented Kenya and Tanzania as elephant savers and Zimbabwe and South Africa as elephant killers. In fact, there were only five African countries in favour of a ban at CITES. The 1989 decision resulted in total polarisation over the ivory issue between East and southern Africa and between southern Africa and the rest of the world. One of the problems with the debates over the ivory ban in the 1980s was that it treated the African elephant as a single population that covered most of the continent. In the post-ban period, Zimbabwe has been keen to point out that different populations of African elephants require different management policies. The experience of 1989 has heavily influenced debates about the ivory trade in the 1990s, and the mistrust and hostility created by the way the ban was instituted in 1989 carried over into subsequent CITES meetings, where pro-banners and anti-banners took increasingly entrenched positions.

The post-ban ivory trade

The main components of the pro-ban argument were that only a ban would effectively halt the trade and that the ban would stop all poaching. Initially, the ban had the desired effect and demand was cut, especially in Western states, where it was easy to appeal to ideas of elephants being slaughtered to pay for Western vanity. This reversal of consumer opinion was one of the most important successes of the campaign to stop the ivory trade (Currey and Moore, 1994: 13). It has certainly had a lasting effect, since it is now considered socially unacceptable to buy or wear ivory. The collapse in demand made enforcement of the trade ban much easier. Poaching also declined, because of the drop in ivory prices and lack of access to legal markets, and this gave elephant populations time to recover. The price received by middlemen in East Africa for ivory dropped to US$4.5 per kilo in 1990, which was a fraction of its pre-ban value (Currey and Moore, 1994: 4). However, this situation did not continue, because bans are fairly blunt and symbolic gestures. In instituting a ban, CITES parties did not consider how it would be enforced at a local level (despite the experience of the failed rhino horn ban) and, in the 1990s, it has become clear that the ban has achieved neither of its objectives as effectively as it was supposed to do, thereby allowing a political space in which pro-trade interest groups can continue to lobby for a reopening of a legal and tightly controlled ivory trade.

At each CITES conference since 1989, there has been a rerun of the level of debate which surrounded the ban in the first place. Over the

1990s, southern African range states became more convinced that the ban was not working and was, in fact, in contravention of their interests. The tenth CITES conference, held in the heated atmosphere of Harare, proved to be the arena where all the main interest groups converged to battle it out over a partial reopening of the ivory trade. The first proposal, to down-list the African elephant to Appendix II, was the most ambitious and politically contentious, but it was agreed amongst pro-trade interest groups that, if the down-listing proposal failed, the second choice was to allow Namibia, Botswana and Zimbabwe only to trade with Japan (as the largest ivory consumer), as long as the four states involved could prove that they had adequate controls to prevent illegal ivory trading. This was the first time that southern Africa had a real chance of overturning the ivory ban since 1989. The concerns of southern African states had to be taken seriously, because of the real possibility that Zimbabwe (in particular) would go it alone and declare a reopening of the ivory trade. Consequently, the same tactics of media hype, blackmail, with threats of withdrawal of aid, rumour and counter-rumour were used by alliances formed from the ever more entrenched 1989 interest groups. This meant that the conference was about for versus against, with little political space for uncertainty.

The animal welfare lobby again played a major part in the 1997 CITES meeting, despite their continuing role as lobbyists rather than voting members. In particular, the International Fund for Animal Welfare (IFAW) openly opposed Zimbabwe's position on the ivory trade. It was part of the Species Survival Network, a coalition of 50 NGOs, which included internationally-famous environmental NGOs, such as Greenpeace, formed in 1992 and dedicated to strict enforcement of CITES regulations. IFAW and its allies argued that Zimbabwe and other southern African states would not be capable of preventing illegal exports if CITES allowed a partial or total lifting of the ivory ban.[26] The management problems and the power struggle in Zimbabwe's Parks Department were also used to full effect as part of the discussions about the ivory ban. A confidential report by a panel of CITES experts noted that illegal ivory trading had continued unchecked in Zimbabwe, citing, for example, a consignment of carving made from 70 tusks that had been shipped to Japan. In addition, critics of the Parks Department's capacity to manage a legal ivory trade pointed out that, since the Department had been systematically starved of funds by the central government, it needed every source of revenue available, including the sale of ivory, and that its lack of management capability made it unlikely that the money would be used effectively for elephant conservation.[27]

[26] *Star* (South Africa) 11.6.97 'Reopening of Trade Slated as Danger to the Big Three'; *Star* 13.6.97 'Massive Illegal Wildlife Trade Revealed'; *Herald* 11.6.97 'Downlisting of Jumbo: IFAW Still Opposed'.
[27] *Mail and Guardian* 6.12.96 'Zim's Illicit Ivory Trade Exposed'; *Mail and Guardian* 31.6.97 'Zim Parks in Trouble'.

The pro-ban alliance also used the debt and aid question to try to force pro-trading states into withdrawing their proposal for a partial lifting of the ivory ban. The southern African states accused the animal rights lobby of engaging in international blackmail and spreading misinformation. It was reported that some observers at the CITES conference believed that members of the pro-ban alliance had paid some African countries to vote against southern Africa, thereby exploiting the historical split between East and southern Africa on the ivory trade. Indeed, it was also reported in the local press at the time of the conference that animal rights lobbyists and Northern states opposed to the trade had threatened to withdraw development aid from states that supported Zimbabwe, Namibia and Botswana, and that Northern delegates at CITES had stated they would only vote if there was a common African policy on the ivory trade.[28] In this way, the debt relief question was captured by animal rights NGOs and other anti-trade interest groups to force African states to promise never to engage in ivory trading.

On the other side, Zimbabwe and its allies in the Far East and the southern African region mobilised all its support; this included rural communities and their commitment to Campfire as the central means of legitimating a reopening of the ivory trade. A major political bargaining chip at CITES was that Zimbabwe held one of the largest and most secure elephant populations on the continent, and the delegates at CITES have always been aware of the importance of keeping such important elephant range states within the boundaries of the Convention. Consequently, during the conference it was intimated, that if CITES members did not take account of southern African interests, Zimbabwe would break the international trade ban. The Minister of Environment and Tourism, Chen Chimutengwende, suggested that, if CITES did not act in favour of Zimbabwe's interests, it would have made itself irrelevant to Zimbabwe.[29] The fact that Zimbabwe has such a large elephant population meant that this was a threat that none of the parties at CITES could ignore, even though it was a risky position for Zimbabwe to take. In response, the anti-trade interest groups accused the Zimbabwean delegates of blackmail and pursuit of selfish economic interests.

Secondly, the pro-trade alliance argued that, although organisations such as EIA have trumpeted the success of the Appendix I listing, the situation is more complex. The pro-trade interest groups argued that the ban had not shut down markets for ivory as effectively as its supporters suggested and that, since there was a continuing demand for ivory across the world, Zimbabwe had the right to utilise its resources to take

---

[28] *Herald* 12.6.97 'African States Divided On The Ivory Issue'; *Herald* 13.6.97 'Maveneke Rejects Debt Relief on Ivory'.

[29] *Herald* 16.6.97 'CITES or No CITES, We'll Go it Alone'; *Pretoria News* (South Africa) 17.6.97 'Zimbabwe Backs Off Threat to Break Ivory Ban'; *Guardian* 17.6.97 'Harare Resists the Ivory Sale Ban'; *Guardian* 18.6.97 'Ivory Trade States to Try Again After Near Miss'.

advantage of such markets. Overall, although ivory remains difficult to sell legally and the markets of the West have been closed, the illegal markets of Asia began to re-establish themselves as early as 1991. In addition, embryonic pre-ban Asian markets have continued to grow in North Korea and South Korea (Dublin and Jachmann, 1991: 62–3).

One of the main arguments for the ban was that it would stop poaching. Predictably, in line with their arguments against the ban in the first place, Zimbabwe continued to argue that poaching rose noticeably after the ban. Research by the IUCN African Elephant Specialist Group suggests that, where poaching did decline in the initial post-ban period, it was largely due to increased expenditure on law enforcement, rather than the ban *per se* (Dublin et al., 1995). Most donor funding was poured into East Africa and it is questionable whether the donor community would have been willing to provide as much finance for every African state as it provided for Kenya. Similarly, seizures of ivory have begun to increase again, with 35 tonnes discovered between 1992 and 1994.[30] This reflects the re-establishment of illegal markets and smuggling routes. The Zimbabwean policy position has continued to be that the ban frustrates elephant conservation in the long term and that Zimbabwe's problem is one of too many elephants rather than too few, leading to increased human–wildlife conflicts. In response, EIA has accused Zimbabwe of a politically motivated campaign of misinformation, which diverts funds away from real conservation to the politics of the ivory ban (EIA, 1992: 22).

The southern African alliance also captured and mobilised the debate surrounding debt relief and aid to argue in favour of a partial end to the ban. The Southern African Development Community (SADC) countries put forward a proposal for a debt for ivory buy-out for African range states that had huge stockpiles but were unwilling to engage in the ivory trade. The proposal was specifically drafted by the southern African pro-trade alliance to allay the fears of East African elephant range states and attempt to weld together a common African elephant policy. Conservationists had been concerned about the growing African ivory stockpiles, because the pressure to trade increased as the stockpiles grew. For example, the Trade Records Analysis of Flora and Fauna in International Commerce (Traffic) research indicated that, by 1995 the total continental stockpile was already 500 tons.[31] There have been repeated calls for a committee of CITES to investigate the issue of stockpiles (Dublin et al., 1995). One possible solution was that the stocks could be bought with a one-off payment from the World Bank's Global Environment Facility (GEF) and then burnt, or that conservation organisations could buy them.[32] In this way, African states would obtain revenue for their resources without a

---

[30] *Economist* 5.11.94 'Whose Elephants Are They Anyway?', p. 70; *Star* 13.6.97 'Massive Illegal Wildlife Trade Revealed'.
[31] *Traffic Dispatches* Sept. 1995, p. 6.
[32] Interview with John Gripper, Sebakwe Black Rhino Trust, 14.10.94, Ascott-Under-Wychwood.

renewed international trade. It was suggested that only ivory stocks outside southern Africa should be bought by Western aid agencies and then the money used to fund elephant conservation.[33]

Finally, the presence of Zimbabwean rural community groups at the conference was used to re-emphasise that ivory trading was as much about rural development as about revenue generation. The director of the Campfire Association, Taparendava Maveneke, stated that holding the conference in Harare was important, because it was the first time that rural communities had come face to face with key policy-makers in the field of international wildlife trading, and that this was significant because it meant parties at CITES would be made to understand the problems being faced by the various parties to the convention.[34] The Parks Department in Zimbabwe and its domestic and international allies have always been aware that Campfire forms the central political justification for continued commitment to sustainable utilisation. During the course of the conference, the Zimbabwean Government tried to convince members that were wavering on the ivory issue by organising weekend visits to Campfire areas.[35] The Campfire Association's vocal backing for a reopening of the ivory trade was a risky position in terms of international support for the programme, since it was feared that Campfire's support for the ivory trade would infuriate the US government, which, on the one hand, supported Campfire, but, on the other, was vigorously opposed to the ivory trade. Environmental NGOs have found the argument that wildlife is a resource to be used to benefit rural communities one of the hardest to reject. Indeed, there were accusations at CITES that environmental NGOs and preservationist Western states were frustrating developing countries by imposing a new form of imperialism in the realm of the environment through their insistence that African states could not be in charge of their own resources. This dispute essentially grew from competing definitions of wildlife as a national resource to be used for domestic benefits, and the view of Western states and international environmental NGOs that elephants were an international resource to be protected at all costs for the global good.

Finally, at the 1997 meeting, Zimbabwe, Namibia and Botswana were successful in having their elephant populations down-listed to Appendix II, with the proviso that trading would only be allowed in 1999 if adequate measures were introduced to ensure that trading could be sustainable and free from illegally hunted ivory. The vote was surrounded by accusations of a lack of transparency and of blackmail. In the first round of voting, the southern Africa proposal to down-list the region's elephants to Appendix II failed to obtain the necessary two-

---

[33] *Pretoria News* 17.6.97 'CITES Proposal to Buy Some of Africa's Ivory Stocks'; *Herald* 13.6.97 'Maveneke Rejects Debt Relief on Ivory Trade'.
[34] *Herald* 13.6.97 'Maveneke Rejects Debt Relief on Ivory Trade'.
[35] *Pretoria News* 13.6.97 'SADC Closes Rank On Elephant, Rhino'.

thirds majority. Consequently, there had to be a second round of voting, which focused on individual country proposals. The SADC countries pressed for a secret ballot, so that voting members could not be black-mailed by the animal welfare lobby or by aid donor countries, such as the USA. In return, the animal welfare lobby accused the SADC countries of making CITES lean towards a lack of transparency and democracy. In the end, the secret ballot meant that Namibia, Zimbabwe and Botswana could reopen a restricted trade with Japan. Of course, the decision was immediately met with the concern that this would provide a signal to poachers to go ahead and restart the levels of commercial poaching witnessed in East Africa in the 1980s.[36] It remains to be seen if the partial lifting of the ivory ban does result in a renewed onslaught on elephants, but undoubtedly the anti-trade interest groups will be regrouping to argue in favour of a renewed ban on ivory trading in the future.

In conclusion, CITES has come to be perceived in southern Africa as an example of an international prohibition regime. In southern Africa, it is reviled as an organisation that is interested in trade bans as a conser-vation method. The difficulty with blanket trade bans is that they cannot possibly take account of local conditions, because, at one stroke, trade bans create an international norm. The debates over the international trade in endangered species reveal major divisions between different factions of the conservation movement. On the one hand, there has been disagreement over the precise meaning of the term 'endangered', and CITES has been concerned to define what constitutes legitimate use of animals. It has moved beyond its initial role as a trade regulator to an organisation interested in conservation. In the context of CITES, the meaning of 'endangered' has been determined by NGO campaigns, which have played a critical part in influencing the direction of CITES. In this way, ideas of animal rights and deep-green ideologies have been globalised through international trade bans.

The trade ban on ivory and rhino horn has deepened divisions between supporters and opposers of sustainable utilisation. The ivory issue resulted in a deep mistrust between East African and southern African states over elephant management. The rift over elephants and rhinos demonstrates the difficulties involved in an international community attempting to regulate what is considered a national resource. It is clear that conservation policy has to operate at a domestic level and within an international context. This conflict of interests has arisen over a number of environmental issues, and the debate centres around whether environ-mental goods such as elephants or rain forests can truly be considered a national resource. Deep-green conservationists and environmentalists argue that elephants and rhinos are too important to be left to the vagaries of national policy, especially when the range states intend to cull, trade

---

[36] *Herald* 18.6.97 'CITES: Three States Down But Not Out'; *Herald* 20.6.97 'At Last, the Ivory Trade War is Won!'; *Mail and Guardian* 27.6.97 'All Clear for Ivory Trade'.

or sport hunt them. Some environmentalists have argued that these species are so symbolically important that they are a world resource, which should be subject to international management and policy decisions. In turn, range states, such as Zimbabwe, view this as a new environmental imperialism, arguing that Zimbabwean people have to live with elephants and bear all the costs of keeping them, and so the state is the only entity that should decide their fate. Moreover, policy-makers in Zimbabwe felt aggrieved that control over policy decisions has been taken away from them and passed to the international community, which they perceived to be dominated by preservationists. It is clear that Zimbabwe's domestic conservation policy has to operate within an international context and that that context has proved to be a significant constraining influence on the operation of sustainable utilisation as a policy.

CONCLUSION

Environmental politics is caught in its own rhetoric, because environmental interest groups seek to present conservation as a policy based on a kind of neutral science that has no social, economic or political impact, and those same interest groups mobilise distinctly political ideologies of the environment to justify their own position and to denounce the stance of their opponents. The 1997 Convention on International Trade in Endangered Species (CITES) conference provided a perfect example of this. CITES Chair, Nobutoshi Akao, opened the conference by stating 'it is not an emotional and single-minded pursuit of a specific philosophical approach to conservation which should prevail here. It is reason, logic and cool-headed judgement, based on scientific data, which should be the basis of our decisions.'[1] This statement assumed that there are such things as reason, logic and an apolitical science in the realm of conservation and yet, of all the interest groups and institutions involved in Zimbabwe's conservation policy, CITES is the most highly politicised, because it spans the local, national and global. The failure to recognise that appeals to reason, logic and neutral science are as much a political ideology of conservation as any other approach to wildlife is one of the worst aspects of environmental politics. CITES, like other decision-making bodies, formulates policies and takes decisions based on political lobbying, which, in turn, relies on emotion, philosophical standpoints and political manoeuvring. Furthermore, the involvement of politics, emotion and ideology in conservation is not necessarily a negative factor. Rather, local people, national governments and international institutions have to find a political accommodation for their conservation decisions, and a situation that all parties can live with can only be reached through political negotiation, not through a supposed apolitical science of biological conservation.

The highly politicised nature of conservation is equally the case in campaigns run by international environmental NGOs to save wildlife in

---

[1] *Pretoria News* 10.6.97 'Clash Looms at CITES Conference'.

173

Africa as in the Zimbabwe Parks Department's use of Communal Areas Management Programme for Indigenous Resources (Campfire) as the centre-piece of a domestic and international political justification of sustainable utilisation. In global environmental debates, the often furious clashes over wildlife are about control of important resources and competing definitions of whom these resources belong to. For instance, are elephants a local, national or global resource, are elephants to be utilised by rural communities that live around them for their own benefit or are they to be preserved in the interests of the common global environmental good? For developing countries, it is essential that conservation policies are socially, economically and politically, as well as environmentally, sustainable. The artificial separation of the human and natural worlds throughout sub-Saharan Africa in the colonial period led to conservation policies that were resisted and were ultimately bound to fail, because they were politically untenable. The exclusionary principle was the basis on which national parks were founded. Protected areas and their associated anti-poaching policies led to an artificial and politically unsustainable separation of humans and wildlife. This resulted in national parks that were constantly compromised by the indigenous peoples, whose rights of access to wildlife had been removed. Subsistence poaching, then, became a form of resistance to wildlife policies that did not have any legitimacy for local people. This resulted in anti-poaching policies that attempted to enforce the separation of indigenous peoples and wildlife, and it was this that motivated poaching, because illegal use of wildlife was the only value that wildlife retained for local people.

Zimbabwe's commitment to sustainable utilisation is presented by the wildlife authorities as an attempt to find a wildlife policy that is politically, economically, socially and environmentally sustainable. Sustainable utilisation arose out of a commitment to a specific policy that was rooted in the country's history and the relative abundance of wildlife. Wildlife did not have a high preservation value, because the large wildlife populations in Southern Rhodesia and later Zimbabwe meant that wildlife has a high utilisation value. Wildlife is considered to be a natural resource, alongside other environmental goods, such as forests, minerals and water, which are all to be utilised to further development. This is an explicit statement of an environmental ideology that places the satisfaction of basic human needs above the rights of animals, and so Zimbabwe's wildlife policy is at odds with environmental ideologies that are based on animal rights or animal liberation. Rather, Zimbabwean policy revolves around notions of human stewardship over nature. The policy is based on the premise that people in a developing country will only accept wildlife if it has an economic value, so that, in order to compete with other forms of land use and to survive, wildlife must pay its way. Wildlife preservation in Zimbabwe is ultimately self-defeating, because the state has been unable to raise the domestic political support for preservation rather than sustainable utilisation, and because the development of anti-poaching initiatives based on inducements rather than coercion has highlighted the failure of the state to provide the effective policing power necessary to make preservation work.

The principle of making wildlife pay is also an attempt to reconcile the potentially competing demands of environmental conservation and development. This is encapsulated in policies that allow rural people and commercial farmers to utilise wildlife for economic gain. Under preservationist systems, land under wildlife only has an economic value as a tourist attraction to the government and safari operators, while for local communities it rarely has any value at all. In the system of sustainable utilisation favoured by Zimbabwe, land under wildlife yields revenue through sport hunting, cropping and wildlife viewing. Conserving wildlife and the habitat that it lives in has become economically advantageous to landowners and rural people. Revenue derived from wildlife can then be used for development projects and contributes to government foreign exchange reserves through tourist spending. The involvement of the private sector in wildlife conservation has become a critical element of Zimbabwe's wildlife policy. Commercial cattle farmers switched to wildlife production, because the policy framework and economic advantages provided powerful incentives for them, resulting in a new kind of wildlife conservation, most significantly for the black rhino, in privately run conservancies. The support for privatised wildlife conservation from the Parks Department reflected its belief that the state sector had proved unable to protect key endangered species.

The state's involvement in the wildlife sector demonstrates the importance of the politics of conservation, particularly the fact that, to be acceptable to government and to local people, wildlife policy has to take account of local political conditions. The Parks Department has sought to devolve authority away from the central state to individual landowners and local communities. Moves to take responsibility away from the state have arisen from concerns that wildlife was fast disappearing outside the Parks and Wildlife Estate and that there was a need to remove wildlife from a position where corrupt elements in the state bureaucracy and ruling party could obtain control over it. This was clear in the power struggle within the Parks Department. The conservationist faction was marginalised and removed from positions of influence in the Parks Department, partly because of its attempts to remove wildlife from the state sector. This reflected the peculiar position of powerful elements within the ZANU-PF, such as Joshua Nkomo, and his influence was felt in the Parks Department through pressure from figures in the Ministry of Environment and Tourism. The conservationist faction was convinced that Nkomo and his allies were keen to gain control over potentially lucrative wildlife and tourism resources, which were under the control of the Parks Department.

The case of the Parks Department also demonstrates the extent to which wildlife is an arena for political competition. Wildlife conservation is a racially controversial area of policy, which has been perceived as a peculiarly white concern, resulting in attacks on the commercial wildlife ranching sector and on the white Parks officers. The leadership of ZANU-PF has been reluctant to be seen to favour the wildlife sector, because it is so closely bound up with the land question in Zimbabwe, and land under wildlife is presented as unproductive, while commercial farms under wildlife are perceived to be perpetuating the racially uneven land distribution. The debates over the role of the Parks Department and the power struggle within it represent a microcosm of certain aspects of Zimbabwean politics, such as racial divisions, the land question, the role of the commercial farming sector and internal struggles in the ruling party.

The policy of sustainable utilisation has been criticised by Western environmental NGOs and donors. The criticism stems from the belief that animals should not be treated as commodities or defined as resources and, for a number of deep-green NGOs, such as the Humane Society, animals have an intrinsic right to life, which is not linked to any value they have for humans. Disagreement has also arisen with some NGOs over allegations of corruption and the involvement of officials and the Zimbabwe National Army in poaching. For example, the Environmental Investigation Agency (EIA) argues that sustainable utilisation is made a nonsense if elements in the state and in the army unsustainably utilise wildlife. The driving forces behind commercial poaching were organised criminal syndicates, armies and wealthy dealers in Europe and Asia. In the 1980s, elephants and rhinos were victims of a by-product of the Cold War, which saw the proliferation of automatic weapons on the continent. Yet anti-poaching policies largely concentrated on the stereotyped poverty-stricken local poacher, rather than on corrupt wildlife services and the armed forces. This disagreement is important to the Parks Department in Zimbabwe because it permeates and defines debates with NGOs and donors about which kinds of funding they are prepared to commit to. NGOs and donors can influence domestic conservation policy in a number of ways, such as provision or denial of funds and through campaigning. This also reflects the growing trend of the privatisation of international relations that has seen the decline of state-to-state relations and the transfer of international relations into the hands of the private sector and NGOs. NGOs and donors have tended to choose to fund projects that closely reflect the demands of their home constituencies or the particular agenda of the organisation, and this has led to charges of eco-imperialism being lodged by Zimbabwe against various conservation NGOs. This was nowhere more apparent than in debates over the rhino horn and ivory trades, where the arguments over the trade in products from endangered species clearly demonstrate the political and ideological nature of conservation policy. If conservation were purely a matter of scientific management, the levels of debate and disagreement over how best to manage elephant and rhino populations would not arise. The clash of interests at CITES highlighted divisions between East African and southern African states and the disagreements between Western states and southern Africa. It also revealed the importance of clearly defining what constitutes a domestic, regional or international resource. The African elephant may be used internationally by conservation organisations to raise funds, but national conservation departments are responsible for the day-to-day management of those elephant populations.

Wildlife conservation in southern Africa has social, political and economic consequences, because it requires the government to set aside resources and land. The land question, in particular, is at the heart of Zimbabwean politics, and any policy that requires land will be instantly political. Sustainable utilisation of wildlife has proved to be the only policy that is workable in Zimbabwe. This is because the state has been unable to enforce preservationist measures, as was clear in the failed anti-poaching initiatives and the switch to local conservation programmes in the form of Campfire. The second factor is that sustainable utilisation enjoys local legitimacy, and so any attempt to implement preservationist measures would be politically, economically and socially untenable. Sustainable utilisation is the only means of ensuring the continued survival of wildlife in Zimbabwe, and possibly throughout sub-Saharan Africa. While the state provides a regulatory framework in which consumptive and non-consumptive utilisation takes place, the international context in which wildlife is used indicates that African wildlife is a matter of international as well as local concern. Consequently, conservation NGOs and CITES provide a degree of external accountability for domestic wildlife conservation agencies.

Despite the attempts to depoliticise conservation with appeals to apolitical science, it is clear that the realm of the environment is highly political. Zimbabwe needs a wildlife policy that is politically acceptable to its domestic constituency, and the pursuit of sustainable utilisation is the Parks Department's response to that need. Sustainable utilisation is about finding a solution to the human versus wildlife conflict that is socially acceptable, politically tenable and environmentally sustainable. Unfortunately, two key problems that threaten the future success of sustainable utilisation are the power struggle in the Parks Department and the failure to tackle poaching, which is carried out by highly organised, armed interest groups throughout the region. It is here that the environmental interest groups that oppose utilisation on moral and political grounds have their strongest argument against Zimbabwe's hard-line stance. However, their own political lobbying against utilisation does not acknowledge the demonstrable failure of preservationist policies, which have proved to be politically untenable and will result in wildlife being removed from Africa slowly but surely, because local people have no reason to conserve animals that have no benefit beyond the aesthetic.

Abel, N.O.J. and P. Blaikie, 1990, *Land Degradation, Stocking Rates and Conservation Policies in the Communal Rangelands of Botswana and Zimbabwe.* London: ODI Pastoral Network Paper 29a.

Adams, W.M., 1990, *Green Development* (London: Routledge).

Adams, W.M., 1992, *Wasting the Rain: Rivers, People and Planning in Africa* (London: Earthscan).

Ades, A. and R. di Tella, 1996, 'The Causes and Consequences of Corruption: A Review of Recent Empirical Contributions' *IDS Bulletin* 27: 6–11.

Agarwal, A., 1995, 'Dismantling the Divide Between Indigenous and Scientific Knowledge' *Development and Change* 26: 413–39.

Ahonsi, B.A., 1995, 'Gender Relations, Demographic Change and the Prospects for Sustainable Development in Africa' *Africa Development* 20: 85–114.

Akama, J.S., C.L. Lant and G. Wesley, 1996, 'A Political Ecology Approach to Wildlife Conservation in Kenya' *Environmental Values* 5: 335–47.

Allen, T. and A. Thomas (eds), 1992, *Poverty and Development in the 1990s* (Milton Keynes: Open University Press).

Amnesty International, 1992, *Zimbabwe: Poaching and Unexplained Deaths: The Case of Captain Nleya* (London: Amnesty International).

Amsden, A., 1990, 'Third World Industrialisation: "Global Fordism" or a New Model?' *New Left Review* 182: 5–32.

Anderson, D. and R. Grove (eds), 1987, *Conservation in Africa* (Cambridge: Cambridge University Press).

Ap, J., 1992, 'Resident's Perceptions on Tourism Impacts' *Annals of Tourism Research* 19: 665–90.

Attfield, R., 1983, 'Christian Attitudes to Nature' *Journal of the History of Ideas* 44: 369–86.

Barbier, E., 1987, 'The Concept of Sustainable Environmental Conservation' *Environmental Conservation* 14: 101–7.

Barbier, E. et al., 1990, *Elephants, Economics and Ivory* (London: Earthscan).

Barbier, E. et al., 1994, *Paradise Lost? The Ecological Economics of Biodiversity* (London: Earthscan).

Barrett, C.B. and P. Arcese, 1995, 'Are Integrated Conservation Development Projects (ICDPs) Sustainable? On Conservation of Large Mammals in Sub-Saharan Africa' *World Development* 23: 1073–84.

Bartlemus, P., 1986, *Environment and Development* (London: Allen and Unwin).

Bates, R.H., 1981, *Markets and States in Tropical Africa* (Berkeley: University of California Press).

Bates, R.H. and M. Lofchie (eds), 1980, *Agricultural Development in Africa* (London: Praeger).

Baxter, J. and J. Eyles, 1997, 'Evaluating Qualitative Research in Social Geography: Establishing "Rigour" in Interview Analysis' *Transactions of the Institute of British Geographers* 22: 505–25.

Bayart, J.F., 1993, *The State in Africa: the Politics of the Belly* (London: Longman).

Beck, U., 1995, *Ecological Politics in an Age of Risk* (Cambridge: Polity Press).

Beckerman, W., 1994, 'Sustainable Development: Is it a Useful Concept?' *Environmental Values* 3: 191–209.

Beinhart, W., 1984, 'Soil Erosion, Conservationism and Ideas about Development: A Southern African Exploration 1900–1960' *Journal of Southern African Studies* 11: 52–83.

Beinhart, W., 1989, 'The Politics of Colonial Conservation' *Journal of Southern African Studies* 15: 143–63.

Bell, R.H.V., 1987, 'Conservation With a Human Face: Conflict and Reconciliation in African Land Use Planning' in Anderson and Grove (eds) *Conservation in Africa* pp.79–101.

Bhebe, N., 1979, *Christianity and Traditional Religion in Western Zimbabwe 1859–1923* (London: Longman).

Bhila, H.H.K., 1982, *Trade and Politics in a Shona Kingdom* (London: Longman).

Blomstrom, M. and B. Hettne, 1984, *Development Theory in Transition – The Dependency Debate and Beyond: Third World Responses* (London: Zed Books).

Bloom, J., 1996, 'A South African Perspective of the Effects of Crime and Violence on the Tourism Industry' in Pizam and Mansfeld (eds) *Tourism, Crime and International Security Issues* pp. 91–102.

Bolin, B., 1994, 'Science and Policy Making' *Ambio* 23: 25–9.

Bond, I., 1993, *The Economics of Wildlife and Land Use in Zimbabwe: An Examination of Current Knowledge and Issues.* WWF MAPS Project Paper No. 36 (Harare: WWF).

Bond, I., 1994, 'The Importance of Sport Hunted Elephants in Campfire in Zimbabwe' *Traffic Bulletin* 14: 117–19.

Bonner, R., 1993, *At the Hand of Man: Peril and Hope for Africa's Wildlife* (London: Simon and Schuster).

Boo, E., 1990, *Ecotourism: The Potentials and the Pitfalls,* Vol. 1 (Washington DC: WWF).

Bookchin, M. et al. (eds), 1991, *Defending the Earth: A Dialogue Between Murray Bookchin and Dave Foreman* (Boston: South End Press).

Booth, D., 1985, 'Marxism and Development Sociology: Interpreting the Impasse' *World Development* 13: 761–87.

Bowler, M., 1995, *Aerial Census of Elephant and Other Large Mammals in North West Matabeleland, Zambezi Valley and Gonarezhou National Park, Zimbabwe Aug.-Oct. 1993* (Harare: Branch of Terrestrial Ecology, DNPWLM).

Brack, D., 1995, 'Balancing Trade and the Environment' *International Affairs* 71: 497–514.

Bradley Martin, E. and C. Bradley Martin, 1982, *Run Rhino Run* (London: Chatto and Windus).

Bramwell, A., 1989, *Ecology in the Twentieth Century: A History* (London & New Haven: Yale University Press).

Bratton, M., 1990, 'Non-Governmental Organisations in Africa: Can They Influence Public Policy?' *Development and Change* 21: 87–119.

Broad, R., 1994, 'The Poor and the Environment: Friends or Foes' *World Development* 22: 811–22.

Brohman, J., 1996, 'New Directions in Tourism for Third World Development' *Annals of Tourism Research* 23: 48–70.

Brookes-Ball, S., 1993, *Wildlife Society of Zimbabwe* (Harare: Wildlife Society).

Bruessow, C., 1995, 'Protected Areas and Neighbouring Communities: Principles for Resource Sharing and Co-management'. Paper presented at the SADC Annual Conference for the Regional Natural Resource Programme, held in Kasane, Botswana, April 1995.

Bruntland, G.H., 1987, *Our Common Future* (Oxford: World Commission on Environment and Development/Oxford University Press).

Bryant, R.L., 1997, 'Beyond the Impasse: The Power of Political Ecology Third World Environmental Research' *Area* 29: 5–19.

Buchan, A.J.C., 1989a, *An Ecological Resource Survey of Chapoto Ward, Guruve District With Reference to the Use of Wildlife.* WWF MAPS Project Paper No. 6 (Harare: WWF).

Buchan, A.J.C., 1989b, *An Ecological Resource Survey of the Gokwe North Proposed Wildlife Utilisation Area.* WWF MAPS Project Paper No. 2 (Harare: WWF).

Bulmer, M., 1993, 'Interviewing and Field Organisation' in Bulmer and Warwick (eds) *Social Research in Developing Countries* pp. 205–19.

Bulmer, M. and D.P. Warwick (eds), 1993, *Social Research in Developing Countries* (London: UCL Press).

Bureau for International Narcotics and Law Enforcement Affairs, 1998a, *Country Certifications: The Certification Process* (Washington DC: US Department of State).

Bureau for International Narcotics and Law Enforcement Affairs, 1998b, *INL Mission Statement* (Washington DC: US Department of State).

Bureau for International Narcotics and Law Enforcement Affairs, 1998c, *International Narcotics Control Strategy Report, 1997: Money Laundering and Financial Crimes* (Washington DC: US Department of State).

Burgess, R.G. (ed.), 1982a, *Field Research: A Sourcebook and Field Manual* (London: Allen and Unwin).

Burgess, R.G., 1982b, 'The Unstructured Interview as a Conversation' in Burgess (ed.) *Field Research* pp. 101–11.

Butler, V., 1995, 'Is This the Way to Save Africa's Wildlife?' *International Wildlife* 25: 38–45.

Caldwell, J.R. and J.G. Barzdo, 1985, *The World Trade in Raw Ivory 1983 and 1984* A Report prepared for the CITES Secretariat (Cambridge: Wildlife Trade Monitoring Unit, IUCN Conservation Monitoring Centre).

Callicott, J.B., 1983, 'Traditional American Indian and Traditional Western European Attitudes Towards Nature: An Overview' in Elliot and Gare (eds) *Environmental Philosophy* pp. 231–59.

Callister, D.J. and T. Bythewood, 1995, *Of Tiger Treatments and Rhino Remedies: Trade in Endangered Species Medicines in Australia and New Zealand* (Sydney: Traffic Oceania).

Care for the Wild, [1992?] *The Elephant Harvest? An Ethical Approach* (West Sussex: CFTW). Introduction by Daphne Sheldrick.

Carruthers, J., 1989, 'Creating a National Park 1910–1926' *Journal of Southern African Studies* 15: 188–217.

Carruthers, J., 1996, 'Defending Kruger's Honor? A Reply to Professor Hennie Grobler' *Journal of Southern African Studies* 22: 473–80.

Carson R., 1962, *Silent Spring* (London: Hamish Hamilton).

Cartwright, J., 1991, 'Is There Hope for Conservation in Africa?' *Journal of Modern African Studies* 29: 335–71.

CASS/WWF/Zimtrust, 1989, *Wildlife Utilisation in Zimbabwe's Communal Lands Collaborative Group Progress Report* (Harare: CASS/WWF/Zimtrust).

Cater, E., 1994, 'Ecotourism in the Third World – Problems and Prospects for Sustainability' in Cater and Lowman (eds) *Ecotourism* pp. 69–86.

Cater, E., 1995, 'Environmental Contradictions in Sustainable Tourism' *Geographical Journal* 161: 21–8.

Cater, E., and G. Lowman (eds), 1994, *Ecotourism: A Sustainable Option* (London: John Wiley and Sons).

Caulfield, H.P., 1989, 'The Conservation and Environmental Movements: An Historical Analysis' in Lester (ed.) *Environmental Politics and Policy* pp. 13–57.

CCTA and IUCN, 1963, *Conservation of Nature and Natural Resources in Modern African States.* A Report on the Symposium Organised by CCTA and IUCN, Arusha, Tanganyika, 1961 (Gland, Switzerland: IUCN).

Central Statistical Office, 1993, *Annual Migration and Tourist Statistics* (Harare: CSO).

Chadwick, D.H., 1992, *The Fate of the Elephant* (London: Penguin).

Chalker, L., 1991, *Good Government and the Aid Programme* (London: ODA).

Chambers, R., 1983, *Rural Development: Putting the Last First* (Harlow, Essex: Longman).

Chambers, R., 1988, 'Sustainable Rural Livelihoods: A Key Strategy for People, Environment and Development' in Conroy and Litvinoff (eds) *The Greening of Aid* pp. 1–18.

Cheater, A., 1990, 'The Ideology of Communal Land Tenure in Zimbabwe: Mythogenesis Enacted?' *Africa* 60: 188–206.

Child, B., 1988, 'The Role of Wildlife Utilisation in the Sustainable Economic Development of Semi Arid Rangelands in Zimbabwe'. Unpublished D.Phil Thesis, Oxford University.

Child, B., 1995a, 'Can Devolved Management Conserve and Develop the Management of Natural Resources in Marginal Rural Economies? The Example of Campfire in Zimbabwe'. Unpublished Paper, DNPWLM, Harare.

Child, B., 1995b, 'Guidelines for the Revenue Distribution Process'. Unpublished policy draft for Campfire, DNPWLM, Harare.

Child, B., 1995c, 'Guidelines for Managing Wildlife Revenues in Communal Lands in Accordance with Policy for Wildlife'. Unpublished policy draft, DNPWLM, Harare.

Child, B., 1995d, *Communal Land Quotas for 1995* (Harare: DNPWLM).

Child, B., 1995e, 'Guidelines for Campfire'. Unpublished policy draft, DNPWLM, Harare.

Child, G., 1984, 'Managing Wildlife for People in Zimbabwe' in McNeely and Miller (eds) *National Parks, Conservation and Development* pp. 118–21.

Child, G., 1995, 'Linkages Between Wild Resource Management and Agriculture in Rural Development' *Campfire Association Publication Series* 1: 27–37.

Child, G. and R. Fothergill, 1962, *Kariba Studies: Techniques Used to Rescue Black Rhinoceros on Lake Kariba, Southern Rhodesia* (Salisbury: National Museums of Southern Rhodesia).

Child, G. and C.R. Savory, 1964, *The Distribution of Large Mammal Species in Southern Rhodesia* (Salisbury: National Museums, Natural Resources Board and DNPWLM).

Chisunga, B. and G. Zirota, 1997, 'Management of the Land and Resources of the Masoka Community of Dande Communal Lands, Zimbabwe' *Society and Natural Resources*, 10: 405–8.

Clapham, C., 1996, *Africa and the International System: The Politics of State Survival* (Cambridge: Cambridge University Press).

Clark, S.L., 1983, 'Gaia and the Forms of Life' in Elliot and Gare (eds) *Environmental Philosophy* pp. 182–98.

Cohen, J.M., 1980, 'Land Tenure and Rural Development in Rural Africa' in Bates and Lofchie (eds) *Agricultural Development in Africa* pp. 349–400.

Cole, M., 1992, 'Private Reserves for Zimbabwe's Thirsty Elephants' *New Scientist* 135: 10.

Conroy, C. and M. Litvinoff (eds), 1988, *The Greening of Aid* (London: Earthscan).

Coolidge, J. and S. Rose-Ackerman, 1997, *High Level Rent Seeking and Corruption in African Regimes: Theory and Cases* (New York: World Bank).

Corbridge, S. (ed.), 1995, *Development Studies: A Reader* (London: Edward Arnold).

Cotgrove, S. and A. Duff, 1980, 'Environmentalism, Middle Class Radicalism and Politics' *Sociological Review* 28: 333–51.

Cousins, B., 1993, *Property and Power in Zimbabwe's Communal Lands: Implications for Agrarian Reform in the 1990s.* NRM Series (Harare: CASS, University of Zimbabwe).

Craig, G.C. and D.StC. Gibson, 1993, *Records of Elephant Hunting Trophies Exported from Zimbabwe* (Harare: DNPWLM).

Cumming, D.H.M., 1990a, *Wildlife Conservation in African Parks: Progress, Problems and Prescriptions.* WWF MAPS Project Paper No. 5 (Harare: WWF/DNPWLM).

Cumming, D.H.M., 1990b, *Wildlife Products and the Marketplace: A View from Southern Africa.* WWF MAPS Project Paper No. 12 (Harare: WWF).

Cumming, D.H.M. and I. Bond, 1991, *Animal Production in Southern Africa: Peasant Practice and Opportunities for Peasant Farmers in Arid Lands.* WWF MAPS Project Paper No. 22 (Harare: WWF).

Cumming, D.H.M., R.F. Du Toit and S.N. Stuart, 1990, *African Elephants and Rhinos: Status Survey and Action Plan.* (Gland, Switzerland: IUCN/SSC African Elephant Specialist Group).

Cunliffe, R.N., 1992, *An Ecological Resource Survey of the Communal Lands of Centenary District.* WWF MAPS Project Paper No. 26 (Harare: WWF).

Cunliffe, R.N., 1994, *The Impact of the Ivory Ban on Illegal Hunting of Elephants in Zimbabwe.* WWF MAPS Project Paper No. 44 (Harare: WWF).

Currey, D. and H. Moore, 1994, *Living Proof: African Elephants: The Success of the CITES Appendix I Ban* (London: EIA).

Daily, G.C., A.H. Ehrlich and P.R. Ehrlich, 1995, 'Socio-economic Equity: A Critical Element in Sustainability' *Ambio* 24: 58–9.

Daly, R.R., 1982, *Selous Scouts Top Secret War, As Told to Peter Stiff,* 3rd edn (Alberton, South Africa: Galago Publishing).

Dann, G., 1996a, 'The People of Tourist Brochures' in Selwyn (ed.) *The Tourist Image* pp. 61–81.

Dann, G.,1996b, *The Language of Tourism: A Sociolinguistic Perspective* (Wallingford, Oxon: CAB International).

Debus, A.G., 1978, *Man and Nature in the Renaissance* (Cambridge: Cambridge University Press).

De La Harpe, D., 1994, 'Wildlife Conservation by the Lowveld Conservancies' *Zimbabwe Hunter* Dec.: 20–2.

De Waal, A., 1997, *Famine Crimes: Politics and the Disaster Relief Industry in Africa* (Oxford: James Currey for the IAI).

Dietz, F.J., U.E. Simonis and J. Van Der Straaten (eds), 1992, *Sustainability and Environmental Policy* (Berlin: Edition Sigma).

Dixon, J. and M. Hufschmidt (eds), 1986, *Economic Valuation Techniques for the Environment* (London: Johns Hopkins Press).

Dixon, J. and P.B. Sherman, 1990, *The Economics of Protected Areas* (London: Earthscan).

DNPWLM, 1991, *Protected Species of Animals and Plants in Zimbabwe* (Harare: DNPWLM).

DNPWLM, 1992a, *Research Plan: DNPWLM, Zimbabwe* (Harare: Branch of Aquatic Ecology and Branch of Terrestrial Ecology, Research Division, DNPWLM).

DNPWLM, 1992b, *Short and Medium Term Action Plans for Black Rhino* (Harare: DNPWLM).

DNPWLM, 1992c, *Zimbabwe Black Rhino Conservation Strategy* (Harare: DNPWLM).

DNPWLM, 1994, *Summary Report on the Campfire Programme and the Campfire Co-ordination Unit* (Harare: DNPWLM).

Dobson, A., 1996, 'Environmental Sustainabilities: An Analysis and Typology' *Environmental Politics* 5: 401–28.

Douglas-Hamilton, I., 1979, *African Elephant Ivory Trade Study: Final Report* (Washington DC: US Fish and Wildlife Service).

Douglas-Hamilton, I., F. Michelmore and A. Indamar, 1992, *African Elephant Database*. EC African Elephant Survey and Conservation Programme (Nairobi: UNEP).

Dovers, S.K., and J.W. Handmer, 1995, 'Ignorance, the Precautionary Principle and Sustainability' *Ambio* 24: 92–7.

Drinkwater, M., 1989, 'Technical Development and Peasant Impoverishment: Land Use Policy in Zimbabwe's Midlands Province' *Journal of Southern African Studies* 15: 287–305.

Drinkwater, M., 1991, *The State and Agrarian Change in Zimbabwe* (London: Macmillan).

Drummond, I. and T.K. Marsden, 1995, 'Regulating Sustainable Development' *Global Environmental Change* 5: 51–63.

Dryzek, J.S and J.P. Lester, 1989, 'Alternative Views of the Environmental Problematic' in Lester (ed.) *Environmental Politics and Policy* pp. 314–30.

Dublin, H.T., 1994a, 'Status and Trends of the African Elephant (*Loxodonta Africana*)'. Paper presented at the African Elephant in the

Context of CITES Conference held in Kasane, Botswana 19–23 September 1994 Department of Environment (DOE) Part B.

Dublin H.T., 1994b, 'A Reassessment of the Impact of the Ivory Ban on the Illegal Hunting of Elephants'. Paper presented at the African Elephant in the Context of CITES Conference held in Kasane, Botswana 19–23 September 1994 Department of Environment (DOE), Part B.

Dublin, H.T. and H. Jachmann, 1991, *The Impact of the Ivory Ban on Illegal Hunting of Elephants in Six Range States in Africa*. WWF International Research Report. WWF Project No. 4578 (Gland, Switzerland: WWF).

Dublin, H.T., T. Milliken and R.F.W. Barnes, 1995, *Four Years After the CITES Ban: Illegal Killing of Elephants, Ivory Trade and Stockpiles* (Gland, Switzerland: IUCN/SSC African Elephant Specialist Group).

Dubos, R., 1986, 'The Wilderness Experience' in Van De Veer and Pierce (eds) *People, Penguins and Plastic Trees* pp. 137–42.

Duffy, R., 1997, 'The Environmental Challenge to the Nation-state: Superparks and National Parks Policy in Zimbabwe' *Journal of Southern African Studies* 23: 441–51.

Duffy, R., 1999, 'The Role and Limitations of State Coercion: Anti-poaching Policies in Zimbabwe' *Journal of Contemporary African Studies* 17: 97–121.

Dunlap, R.E., 1989, 'Public Opinion and Environmental Policy' in Lester (ed.) *Environmental Politics and Policy* pp. 87–134.

Dunlap, R.E. and K. Van Liere, 1978, 'The New Environmental Paradigm: A Proposed Measuring Instrument and Preliminary Results' *Journal of Environmental Education* 9: 10–19.

Dunlap, R.E. and K. Van Liere, 1984, 'Commitment to the Dominant Social Paradigm and Concern for Environmental Quality' *Social Science Quarterly* 65: 1013–28.

Dzingirai, V., 1994, *Politics and Ideology in Human Settlement: Getting Settled in the Sikomena Area of Chief Dobola*. CASS Occasional Paper NRM Series (Harare: CASS).

Eckersley, R., 1992, *Environmentalism and Political Theory: Towards an Ecocentric Approach* (London: UCL Press).

Eden, S.E., 1994, 'Using Sustainable Development: The Business Case' *Global Environmental Change* 4: 160–7.

EIA, 1989, *A System of Extinction: the African Elephant Disaster* (London: EIA).

EIA, 1992, *Under Fire: Elephants in the Front Line* (EIA: London).

EIA, 1994, *CITES: Enforcement Not Extinction* (London: EIA).

EIA, David Shepherd Foundation and Tusk Force, 1993, *Taiwan Kills Rhinos With Your Money: Why You Should Boycott Goods Made in Taiwan* (London: EIA).

Ellenberg, L. et al, 1993, *Community Oriented Wildlife Conservation Programme, Zimbabwe: Mid-term Review Mission* (Germany: GTZ).

Elliot, R. and A. Gare (eds), 1983, *Environmental Philosophy* (Milton Keynes: Open University Press).

Ellis, S., 1989, 'Tuning in to Pavement Radio' *African Affairs* 88: 321–30.

Ellis, S., 1994, 'Of Elephants and Men: Politics and Nature Conservation in South Africa' *Journal of Southern African Studies* 20: 53–69.

Elson, D. (1991) *Male Bias in the Development Process* (Manchester: Manchester University Press).

Falloux F. and L.M. Talbot (1993) *Crisis and Opportunity: Environment and Development in Africa* (London: Earthscan).

Feron, E.M., 1993, 'Mission Report: Participation in International Wildlife Management Congress: Integrating People and Wildlife for a Sustainable Future'. Paper presented in Costa Rica, 19–25 September.

Foote-Whyte, W., 1982, 'Interviewing in Field Research' in Burgess (ed.) *Field Research* pp. 111–22.

Gan, L., 1993, 'The Making of the Global Environmental Facility: An Actor's Perspective' *Global Environmental Change* 3: 256–75.

Geertz, C., 1973, *The Interpretation of Cultures* (New York: Basic Books).

Gelbard, R.S., 1996a, 'The Globalisation of the Drug Trade'. Address at the John Jay College of Criminal Justice, Dublin, Ireland, 20.6.96.

Gelbard, R.S., 1996b, 'The Threat of Transnational Organised Crime and Illicit Narcotics'. Statement before the UN General Assembly, New York City, 21.10.97.

Gellner, E., 1973, *Cause and Meaning in the Social Sciences* (London: Routledge and Kegan Paul).

Ghai, D., 1994, 'Environment, Livelihood and Empowerment' *Development and Change* 25: 1–13.

Gibson, C.C., 1999, *Politicians and Poachers: The Political Economy of Wildlife Policy in Africa* (Cambridge: Cambridge University Press).

Gilbin, J., 1990, 'Trypanosomiasis Control in African History: An Evaded Issue?' *Journal of African History* 31: 59–80.

Goodin, R., 1983, 'Ethical Principles for Environmental Protection' in Elliot and Gare (eds) *Environmental Philosophy* pp. 3–21.

Gorz, A., 1980, *Ecology as Politics* (London: Billing and Sons).

Government of Zimbabwe, 1994a, 'Budget Statement 1994', presented to Parliament 28 July by Acting Senior Minister of Finance E.D. Mnangagwa, MP.

Government of Zimbabwe, 1994b, *Report of the Commission of Inquiry into Appropriate Agricultural Land Tenure Systems* under Chairmanship of Prof. M. Rukuni (Harare: Government of Zimbabwe).

Grindle, M. and J. Thomas, 1989, 'Policy Makers, Policy Choices and Policy Outcomes' *Policy Sciences* 22: 213–48.

Grobler, H., 1996, 'Dissecting the Kruger Myth with Blunt Instruments: A Rebuttal of Jane Carruthers's View' *Journal of Southern African Studies* 22: 455–472.

Grove, R., 1989, 'Scottish Missionaries, Evangelical Discourses and the Origins of Conservation Thinking in Southern Africa 1820–1900' *Journal of Southern African Studies* 15: 163–88.

Grove, R., 1990, 'Colonial Conservation, Ecological Hegemony and Popular Resistance: Towards a Global Synthesis' in MacKenzie (ed.) *Imperialism and the Natural World* pp. 15–51.

Grove, W., 1995, 'The Drug Trade as a National and International Security Threat', paper presented at the Conference Security 94 at the Institute of Strategic Studies and the Security Association of South Africa, University of Pretoria, South Africa.

Hall, C.M., 1994, *Tourism and Politics: Policy, Power and Place* (Chichester: John Wiley and Sons).

Hall, C.M. and V. O'Sullivan, 1996, 'Tourism, Political Stability and Violence' in Pizam and Mansfeld (eds) *Tourism, Crime and International Security Issues* pp. 106–22.

Hall, C.M. and B. Weiler, 1992, 'What's Special About Special Interest Tourism?' in Weiler and Hall (eds) *Special Interest Tourism* pp. 1–14.

Hampton, M.P., 1996, 'Where Currents Meet: The Offshore Interface Between Corruption, Offshore Finance Centres and Economic Development' *IDS Bulletin* 27: 78–87.

Hanlon, J., 1986, *Beggar Your Neighbours: Apartheid Power in Southern Africa* (London: CIIR).

Hanlon, J., 1996, *Peace Without Profit: How the IMF Blocks Rebuilding in Mozambique* (Oxford: James Currey for the IAI).

Hannah, L., 1992, *African People, African Parks: An Evaluation of Development Initiatives as a Means of Improving Protected Area Conservation in Africa* (Washington DC: USAID/Biodiversity Support Programme/Conservation International).

Hardin, G., 1968, 'The Tragedy of the Commons' *Science* 162: 1243–8.

Hargrove, E. (ed.), 1992, *The Animal Rights/Environmental Ethics Debate* (New York: State University of New York Press).

Harrison, D. (ed.), 1992a, *Tourism and the Less Developed Countries* (London: Belhaven Press).

Harrison, D., 1992b, 'International Tourism and the Less Developed Countries: The Background' in Harrison (ed.) *Tourism and the Less Developed Countries* pp. 1–18.

Harrison, D., 1992c, 'Tourism to Less Developed Countries: The Social Consequences' in Harrison (ed.) *Tourism and the Less Developed Countries* pp. 19–34.

Harrison, P., 1992, *The Third Revolution: Population, Environment and a Sustainable World* (London: Penguin).

Hasler, R., 1990, *The Political and Socio-economic Dynamics of Natural Resource Management – Campfire in Chapoto Ward 1989–1990* (Harare: CASS).

Hayter, T., 1989, *Exploited Earth* (London: Earthscan).

Heath, R.A., 1992, 'Wildlife Based Tourism in Zimbabwe: An Outline of its Development and Future Policy Options' *Geographical Journal of Zimbabwe* 23: 59–71.

Hecht, S. and A. Cockburn, 1990, *The Fate of the Forest* (London: Penguin).

Herbert, H.J. and B. Austen, 1972, 'Past and Present Distribution of the Black and Square Lipped Rhinoceros in the Wankie National Park' *Arnoldia* 5.

Herbst, J., 1990, *State Politics in Zimbabwe* (Berkeley: University of California Press).

Higgs, D., 1993, 'The Elephant Men' *Care for the Wild News* 6: 16–17.

Higgs, D., 1994, 'Operation Elevacuation' *Care for the Wild News* 6: 17–24.

Hill, K., 1994, 'Politicians, Farmers and Ecologists: Commercial Wildlife Ranching and the Politics of Land in Zimbabwe' *Journal of African-American Studies* 29: 226–47.

Hill, K., 1996, 'Zimbabwe's Wildlife Utilisation Programmes: Grassroots Democracy or an Extension of State Power' *African Studies Review* 39: 103–21.

Hoare, R.E. and C.S. Mackie, 1993, *Problem Animal Assessment and the Use of Fences to Manage Wildlife in the Communal Lands of Zimbabwe.* WWF MAPS Project Paper No 39 (Harare: WWF).

Hobsbawm, E. and T.O. Ranger (eds), 1983, *The Invention of Tradition* (Cambridge: Cambridge University Press).

Holst, R., 1993, 'Continuity and Change: the Dynamics of Social and Productive Relations in Hwange, Zimbabwe 1870–1960'. Unpublished M.Phil. thesis: University of Liverpool.

Hole, H. M., 1928, *Old Rhodesian Days* (reprint: 1968: London: Frank Cass).

Huntington, S., 1968, *Political Order in Changing Societies* (New Haven and London: Yale University Press).

Hyden, G., 1990, 'Reciprocity and Governance in Africa' in Wunch and Olowu (eds) *The Failure of the Centralised State* pp. 245–69.

Ingram, H.M. and D.E. Mann, 1989, 'Interest Groups and Environmental Policy' in Lester (ed.) *Environmental Politics and Policy* pp. 135–57.

IUCN, 1980, *The International Trade in Rhinoceros Products.* Report Prepared by E.B. Martin for IUCN and WWF (Gland, Switzerland: IUCN/WWF).

IUCN, 1992a, *Rhino Global Action Plan Workshop.* Briefing Book, Gland Switzerland: IUCN Captive Breeding Specialist Group.

IUCN, 1992b, *Analysis of Proposals to Amend the CITES Appendices.* Prepared by the IUCN Species Survival Commission Trade Specialist Group, Traffic and WCMC for the 8th Meeting of the Conference of the Parties, Kyoto, Japan, 2–13 March 1992 (Gland, Switzerland: IUCN).

Ivory Trade Review Group, 1989, *The Ivory Trade and the Future of the African Elephant,* Vol. I, *Summary and Conclusions.* Report prepared for the 7th CITES, Conference of the Parties, Lausanne, October 1989 (ITRG: Oxford).

Jacks, G.V. and R.O. Whyte, 1939, *The Rape of the Earth: A World Survey of Soil Erosion* (London: Faber and Faber).

Jacob, M., 1994, 'Toward a Methodological Critique of Sustainable Development' *Journal of Developing Areas* 28: 237–52.

Jansen, D.J., 1990, *Sustainable Wildlife Utilisation in the Zambezi Valley of Zimbabwe: Economic, Ecological and Political Trade Offs*. WWF MAPS Project Paper No. 10 (Harare: WWF).

Jansen, D.J., undated, *What is a Joint Venture? Guidelines for District Councils with Appropriate Authority*. WWF MAPS Project Paper No. 16 (Harare: WWF).

Johnson, V. and R. Nurick, 1995, 'Behind the Headlines: The Ethics of the Population and Environment Debate' *International Affairs* 71: 547–65.

Jones, M.A., 1991a, *Aerial Census of Elephant and other Large Mammals in the Gonarezhou National Park and Adjacent Areas* (Harare: Branch of Terrestrial Ecology, DNPWLM).

Jones, M.A., 1991b, *Aerial Census of Elephant and Other Large Mammals in North-West Matabeleland Sept.–Oct. 1991* (Harare: Branch of Terrestrial Ecology, DNPWLM).

Jones, M.A., 1994, *Safari Operations in Communal Areas in Matabeleland*. Proceedings of the Natural Resources Management Project Seminar and Workshop Harare: Branch of Terrestrial Ecology, DNPWLM).

Kangwana, K., 1994, 'Human–Elephant Conflict in Relation to CITES: An East African Perspective'. Paper presented at *the African Elephant in the Context of CITES* Conference held in Kasane, Botswana 19–23 September 1994, Department of Environment (DOE), Part B.

Kasere, S., 1995a, 'Campfire: Zimbabwe's Tradition of Caring' *Campfire Assocation Publication Series* 1: 6–17.

Kasere, S., 1995b, 'International Environmental Law: Human Rights or Animal Rights?' *Campfire Assocation Publication Series* 1: 37–46.

Kemf, E. and P. Jackson, 1994, *Wanted Alive: Rhinos in the Wild*. WWF Species Status Report (Gland, Switzerland: WWF-International).

Kenworthy, J., 1987, 'A Historical Review of Tourism and Wildlife Conservation in East, Central and South Africa: The Environmental Impact' in J.C. Stone (ed.) *The Exploitation of Animals in Africa*. Proceedings of a Colloquium at the University of Aberdeen pp. 245–260.

Kenya Wildlife Services, 1996, *Wildlife–Human Conflicts in Kenya* (Nairobi: KWS).

Kepe, T., 1997, 'Communities, Entitlements and Nature Reserves: The Case of the Wild Coast, South Africa' *IDS Bulletin* 28: 47–58.

Kerr, M., 1978, 'Reproduction of Elephant in the Mana Pools National Park, Rhodesia' *Arnoldia* 8.

Kerr, M. and J.A. Fraser, 1975, 'Distribution of Elephant in a Part of the Zambezi Valley, Rhodesia' *Arnoldia* 7.

Khan, M.H., 1996, 'A Typology of Corrupt Transactions in Developing Countries' *IDS Bulletin* 27: 12–21.

King, L.A., 1994, *Inter-organisational Dynamics in Natural Resource Management: A Study of Campfire Implementation in Zimbabwe* (Harare: CASS, University of Zimbabwe).

Kjekshus, H., 1977, *Ecological Control and Economic Development in East African History* (London: Heinemann).

Lal, D., 1995, 'Ecofundamentalism' *International Affairs* 71: 515–28.

Lan, D., 1985, *Guns and Rain: Guerrillas and Spirit Mediums in Zimbabwe* (London & Berkeley: James Currey & University of California Press).

Leach, M., and R. Mearns (eds), 1996, *The Lie of the Land: Challenging Received Wisdom on the African Environment* (Oxford: James Currey for the IAI).

Leach, M., R. Mearns and I. Scoones, 1997, 'Institutions, Consensus and Conflict: Implications for Policy and Practice' *IDS Bulletin* 28: 90–5.

Leader-Williams, N., 1992, *The World Trade in Rhino Horn: A Review* (Cambridge: Traffic International).

Leader-Williams, N., 1994a, 'Sustainable Use of Elephants, with a Focus on East Africa'. Paper presented at the African Elephant in the Context of CITES Conference held in Kasane, Botswana, 19–23 September 1994 (DOE), Part B.

Leader-Williams, N., 1994b, *Evaluation of the Rhino Conservancy Project in Zimbabwe.* WWF MAPS Project Paper No. 43 (Harare: WWF and Beit Trust).

Lemarchand, R., 1988, 'The State, the Parallel Economy and the Changing Structure of Patronage Systems' in Rothchild and Chazan (eds) *The Precarious Balance: State and Society in Africa* pp. 149–70.

Leopold, A., 1986, 'The Land Ethic' in Van De Veer and Pierce (eds) *People, Penguins and Plastic Trees* pp. 73–82.

Lester, J. (ed.), 1989, *Environmental Politics and Policy: Theories and Evidence* (Durham: Durham University Press).

Lindsay, W.K., 1987, 'Integrating Parks and Pastoralists: Some Lessons from Amboseli' in Anderson and Grove (eds) *Conservation in Africa* pp. 149–67.

Lopes, C., 1996, *Balancing Rocks: Environment and Development in Zimbabwe* (Harare: UNDP/SAPES Books).

Lovelock, J., 1979, *Gaia* (Oxford: Oxford University Press).

Lovelock, J., 1988, *The Ages of Gaia* (Oxford: Oxford University Press).

Lugard, F.D., 1893, *The Rise of Our East African Empire,* Vol. 1 (reprint 1968: London: Frank Cass).

Lynn, W., 1992, 'Tourism in the People's Interest' *Community Development Journal* 27: 371–7.

Lyster, S., 1985, *International Wildlife Law* (Cambridge: Grotius Publications).

McIntosh, R., 1996, 'The Emperor Has No Clothes...Let Us Paint Our Loincloths Rainbow: A Classical and Feminist Critique of Contemporary Science Policy' *Environmental Values* 5: 3–30.

MacKenzie, J.M., 1988, *Empire of Nature: Hunting Conservation and British Imperialism* (Manchester: Manchester University Press).

MacKenzie, J.M. (ed.) 1990a, *Imperialism and the Natural World* (Manchester: Manchester University Press).

MacKenzie, J.M., 1990b, 'Experts and Amateurs: Tsetse, Nagana and Sleeping Sickness in East and Central Africa' in MacKenzie (ed.) *Imperialism and the Natural World* pp. 187–212.

McNeely, J. and K. Miller (eds), 1984, *National Parks, Conservation and Development* (Washington: Smithsonian Institution Press).

McNeely, J. et al, 1990, *Conserving the World's Biological Diversity* (Gland Switzerland: IUCN).

Madzudzo, E., 1995a, 'Local Institutions for Natural Resource Management in Bulilimamangwe, Zimbabwe'. Paper presented at Reinventing the Commons, the Fifth Annual Conference of the International Association for the Study of Common Property, held in Bodoe, Norway, 24–28 May 1995.

Madzudzo, E., 1995b, 'Cattle, Grazing and Rangeland Tenure in Bulilimamangwe'. Paper presented at Reinventing the Commons, the Fifth Annual Conference of the International Association for the Study of Common Property, held in Bodoe, Norway, 24–28 May 1995.

Makumbe, J., 1994, 'Bureaucratic Corruption in Zimbabwe: Causes and Magnitude of the Problem' *Africa Development* 19: 45–60.

Malthus, T., 1872, *Essay on the Principle of Population* (reprint 1970: London: Penguin).

Markandya, A. and C. Costanza, 1993, *Environmental Accounting: A Review of the Current Debate.* Environmental Economics Series Paper No. 8 (Nairobi: UNEP Environment and Economics Unit).

Martin, R., 1986, *Communal Areas Management Programme for Indigenous Resources (CAMPFIRE) Revised Version.* Campfire Working Document No. 1/86 (Harare: Branch of Terrestrial Ecology, DNPWLM).

Martin, R.B., 1993, 'Should Wildlife Pay Its Way?' Keith Roby Address, Perth, Australia.

Martin, R.B., 1994a, 'Alternative Approaches to Sustainable Use'. Paper presented at conference on Conservation through Sustainable Use of Wildlife, held in University of Queensland, 8–11 February 1994.

Martin, R.B., 1994b, *The Influence of Governance on Conservation and Wildlife Utilisation: Alternative Approaches to Sustainable Use: What Does and Doesn't Work* (Harare: DNPWLM).

Martin, R.B. and S. Thomas, 1991, *Quotas for Sustainable Wildlife Utilisation in Communal Lands: A Manual for District Councils with Appropriate Authority* (Harare: Zimbabwe Trust/DNPWLM).

Martin, R.B., J.R. Caldwell and J.G. Barzdo, 1986, *African Elephants, CITES and the Ivory Trade* (Gland, Switzerland: CITES).

Martin, R. B., G.C. Craig and V.R. Booth, 1989, *Elephant Management in Zimbabwe: A Review Compiled by DNPWLM* (Harare: DNPWLM).

Martin, R.B., C.A.M. Attwell and M. Rukuni, 1992, *Project Progress Review.* Jan 1992. WWF MAPS Project Paper No. 25, WWF Project No. 3749 (Harare, Zimbabwe: WWF).

Matthews, H.G. and L.K. Richter, 1991, 'Political Science and Tourism' *Annals of Tourism Research* 18: 120–35.

Matose, F., 1997, 'Conflicts Around Forest Reserves in Zimbabwe: What Prospects for Community Management?' *IDS Bulletin* 28: 69–78.

Matzke, G.E. and N. Nabane, 1996, 'Outcomes of a Community Controlled Wildlife Utilisation Program in a Zambezi Valley Community' *Human Ecology* 24: 65–85.

Meadows, D.H., 1972, *The Limits to Growth* (London: Pan).

Metcalfe, S., 1992a, *Community Natural Resource Management: How Non-governmental Organisations Can Support Co-management Conservation and Development Strategies Between Government and the Public* (Harare: Zimbabwe Trust).

Metcalfe, S., 1992b, *Planning for Wildlife in an African Savanna: A Strategy Based on the Zimbabwean Experience: Emphasising Communities and Parks* (Harare: Zimbabwe Trust).

Metcalfe, S., 1992c, *The Campfire Programme in Zimbabwe: 'Empowerment' Versus 'Participation' in Natural Resource Management in the Masoka Community* (Harare: Zimbabwe Trust).

Metcalfe, S., 1996a, *Recommendations from the Working Groups of the Workshop on the Save Valley Conservancy Trust, 28–29 March* (Chiredzi: SVC Trust).

Metcalfe, S., 1996b, 'Benefits of the Save Valley Conservancy to Local Rural People'. Paper presented to the Workshop on the SVC Trust, 28–29 March, Tambuti Lodge, Chiredzi, Zimbabwe.

Metcalfe, S., 1996c, *Enhancing Capacity For Neighbouring Communities to Benefit From the Proposed Save Valley Conservancy Trust.* First Draft Project Identification Plan prepared for RDCs surrounding the Conservancy: Chiredzi, Chipinge, Bikita, Zaka, Buhera (Chiredzi: SVC Trust/Zimbabwe Trust).

Meyer, C.A., 1992, 'A Step Back as Donors Shift Institution Building From the Public to the "Private' Sector" *World Development* 20: 1115–26.

Meyer, C.A., 1995, 'Opportunism and NGOs: Entrepreneurship and Green North–South Transfers' *World Development* 23: 1277–89.

Migdal, J.S., 1988, *Strong Societies and Weak States* (Princeton: Princeton University Press).

Mill, J.S., 1863, *Utilitarianism* (reprinted 1987: London: Everyman).

Milliken, T., 1994a, 'Current Trade in, Demand for and Stockpiles of Ivory in Range and Consumer States'. Paper presented at the African Elephant in the Context of CITES Conference held in Kasane, Botswana, 19–23 September 1994 (Department of Environment (DOE), Part B.

Milliken, T., 1994b, 'A Preliminary Assessment of the Elephant Hide Trade from Southern Africa'. Paper presented at the African Elephant in the Context of CITES Conference held in Kasane, Botswana, 19–23 September 1994 (Department of Environment (DOE), Part B.

Milliken, T., K. Nowell and J.B. Thomsen, 1993, *The Decline of the Black Rhino in Zimbabwe: Implications for Future Rhino Conservation.* (Cambridge: Traffic International).

Mills, J.A., 1993, *Market Under Cover: The Rhinoceros Horn Trade in South Korea* (Cambridge: Traffic International).

Ministry of Environment and Tourism, 1992, *Policy for Wildlife in Zimbabwe* (Harare: Ministry of Environment and Tourism and DNPWLM).

Ministry of Information Immigration and Tourism, 1965, *Report of the Secretary for Information, Immigration and Tourism* (Salisbury: Government Printers).

Ministry of Information, Immigration and Tourism, 1971, *Report of the Secretary for Information, Immigration and Tourism* (Salisbury: Government Printer).

Ministry of Internal Affairs, 1966, *Report of the Secretary for Internal Affairs* (Salisbury: Government Printer).

Ministry of Natural Resources, 1990, *National Conservation Strategy* (Harare: Ministry of Natural Resources).

Momtaz, D., 1996, 'The United Nations and Protection of the Environment: From Stockholm to Rio de Janeiro' *Political Geography* 15: 261–71.

Monbiot, G., 1994, 'Lost Outside the Wilderness' *BBC Wildlife* 12: 28–31.

Moore, D.B., 1991, 'The Ideological Formation of the Zimbabwean Ruling Class' *Journal of Southern African Studies* 17: 472–95.

Moyo, J.N., 1992, 'State Politics and Social Domination in Zimbabwe' *Journal of Modern African Studies* 30: 305–30.

Moyo, S., 1991, *Zimbabwe's Environmental Dilemma: Balancing Resource Inequities* (Harare: Zero Books).

Munt, I., 1994a, 'The "Other" Postmodern Tourism: Culture, Travel and the New Middle Classes' *Theory, Culture and Society* 11: 101–23.

Munt, I., 1994b, 'Ecotourism or Egotourism?' *Race and Class* 36: 49–60.

Murombedzi, J., 1990, *The Need for Appropriate Local Level Common Property Resource Management Institutions in Communal Tenure Regimes* (Harare: CASS, University of Zimbabwe).

Murombedzi, J.C., 1992, *Decentralisation or Recentralisation? Implementing Campfire in the Omay Communal Lands of Nyaminyami District.* CASS Natural Resource Management Working Paper No. 2 (Harare: CASS, University of Zimbabwe).

Murphree, M.W., 1991, 'Communities as Institutions for Resource Management', unpublished paper, CASS Occasional Paper Series (Harare: CASS, University of Zimbabwe).

Murphree, M.W., 1992, *Ivory Production and Sales in Zimbabwe* (Harare: Branch of Terrestrial Ecology, DNPWLM).

Murphree M., 1995, 'Optimal Principles and Pragmatic Strategies: Creating an Enabling Politico-legal Environment for Community Based Natural Resource Management (CBNRM)'. Keynote Address to the

Conference of the Natural Resources Management Programme (SADC Technical Coordination Unit, Malawi, USAID-NRMP Regional) Chobe, Botswana, 3 April 1995.

Murphree, M. and D.H.M. Cumming, 1991, 'Savanna Land Use: Policy and Practice in Zimbabwe'. Paper presented at the IUBS UNESCO/UNEP Conference Workshop Economic Driving Forces and Constraints on Savannah Land Use, Nairobi, Kenya.

Murray, M., 1995, '"Blackbirding" at "Crooks Corner": Illicit Labour Recruiting in the Northeastern Transvaal, 1910–1940' *Journal of Southern African Studies* 21: 373–98.

Mutwira, R., 1989, 'Southern Rhodesian Wildlife Policy (1890–1953): A Question of Condoning Game Slaughter?' *Journal of Southern African Studies* 15: 250–62.

Myers, N., 1995, 'Population and Biodiversity' *Ambio* 24: 56–7.

Myerson, G., and Y. Rydin, 1996, *The Language of the Environment: A New Rhetoric* (London: UCL Press).

Nabane, N., 1994, *A Gender Analysis of Community Based Wildlife Utilisation Initiative in Zimbabwe's Zambezi Valley* (Harare: CASS, University of Zimbabwe).

Nabane, N. and G. Matzke, 1997, 'A Gender Sensitive Analysis of a Community-based Wildlife Utilisation Initiative in Zimbabwe's Zambezi Valley' *Society and Natural Resources* 10: 519–35.

Nadelman, E.A., 1990, 'Global Prohibition Regimes: The Evolution of Norms in International Society' *International Organisation* 44: 479–526.

Naldi, G.J., 1993, 'Land Reform in Zimbabwe: Some Legal Aspects' *Journal of Modern African Studies* 31: 585–600.

Nash, S. [1994?] *Making CITES Work* (Godalming, Surrey: WWF-UK).

Neumann, R.P., 1996, 'Dukes, Earls and Ersatz Edens: Aristocratic Nature Preservationists in Colonial Africa' *Environment and Planning D: Society and Space* 14: 79–98.

Neumann, R.P., 1997, 'Primitive Ideas: Protected Area Buffer Zones and the Politics of Land in Africa' *Development and Change* 28: 559–82.

Norton, A., 1996, 'Experiencing Nature: The Reproduction of Environmental Discourse through Safari Tourism in East Africa' *Geoforum* 27: 355–73.

Norton, G., 1994, 'The Vulnerable Voyager: New Threats for Tourism' *World Today* 50: 237–9.

Nowell, K., Chyi Wei-Lien and Pei Chia-Jai, 1992, *The Horns of a Dilemma: the Market for Rhino Horn in Taiwan* (Cambridge: Traffic International).

Nyamapfene, K.W., 1989, 'Adaptation to Marginal Land Amongst the Peasant Farmers of Zimbabwe' *Journal of Southern African Studies* 15: 384–92.

ODI, 1992, *Aid and Political Reform*. ODI Briefing Paper (London: ODI).

Olindo, P., 1974, 'Park Values, Changes and Problems in Developing Countries'. Paper presented to the Second World Conference on National Parks (IUCN). pp. 52–60.

Olsen, M., D.G. Lodwick and R. Dunlap, 1992, *Viewing the World Ecologically* (Boulder, Colorado: Westview Press).

Oluwo, D., 1990, 'The Failure of Current Decentralisation Programmes in Africa' in Wunch and Olowu (eds) *The Failure of the Centralised State* pp. 74–99.

O'Neill, K.M., 1995, 'The International Politics of National Parks' *Human Ecology* 24: 521–39.

O'Riordan, T., 1988, 'The Politics of Sustainability' in Turner (ed.) *Sustainable Environmental Management* pp.29–51.

O'Riordan, T. and A. Jordan, 1995, 'The Precautionary Principle in Contemporary Environmental Politics' *Environmental Values* 4: 191–212.

Palmer, R., 1977, *Land and Racial Domination in Rhodesia* (London: Heinemann).

Parker, I., 1979, *The Ivory Trade,* Vol. 3: *Discussions and Recommendations* (Nairobi: Wildlife Services Limited).

Peace Parks Foundation, [1997?] *The Origin, Objectives and Activities of the Peace Parks Foundation* (Somerset West, South Africa: Peace Parks Foundation).

Pearce, F., 1991, *Green Warriors: The People and the Politics Behind the Environmental Revolution* (London: The Bodley Head).

Pearce, D., E. Barbier and A. Markandya, 1990, *Sustainable Development* (Aldershot: Edward Elgar).

Peluso, N.L., 1993, 'Coercing Conservation? The Politics of State Resource Control' *Global Environmental Change* 3: 199–217.

Pepper, D., 1984, *The Roots of Modern Environmentalism* (London: Croom Helm).

Peterson, M.J. and T.R. Peterson, 1996, 'Ecology: Scientific, Deep and Feminist?' *Environmental Values* 5: 123–46.

Phimister, I., 1986, 'Discourse and the Discipline of Historical Context: Conservationism and Ideas About Development in Southern Rhodesia' *Journal of Southern African Studies* 12: 262–75.

Phimister, I., 1993, 'Rethinking Reserves: Southern Rhodesia's Land Husbandry Act Reviewed' *Journal of Southern African Studies* 19 (2) 235–9.

Pinchin, A., 1993, *Conservation and Wildlife Management in Zimbabwe* (Bristol: University of Bristol Print Services).

Pitman, D., 1990, 'Wildlife as a Crop' *Ceres* 22: 30–5.

Pizam, A. and Y. Mansfeld (eds), 1996, *Tourism, Crime and International Security Issues* (Chichester and New York: John Wiley and Sons).

Popkin, S.L., 1979, *The Rational Peasant: The Political Economy of Rural Society in Vietnam* (Berkeley: University of California Press).

Porritt, J., 1984, *Seeing Green* (Oxford: Basil Blackwell).

Porter, M., 1990, *The Competitive Advantage of Nations* (London: Macmillan).

Price Waterhouse, 1994, *The Lowveld Conservancies: New Opportunities for Productive and Sustainable Land Use* (Harare: Save Valley, Bubiana and Chiredzi River Conservancies).

Princen, T. and M. Finger, 1994, *Environmental NGOs in World Politics: Linking the Global and the Local* (London and New York: Routledge).

Prins, H.H.T., 1987, 'Nature Conservation as Part of Optimal Land Use in East Africa: The Case of the Maasai Ecosystem in Northern Tanzania' *Biological Conservation* 40: 141–61.

Rabinow, P. and W.M. Sullivan (eds), 1979, *Interpretive Social Science* (Berkeley: University of California Press)

Ranger, T.O., 1985, 'Promises, Promises: An Alternative History of the Rhodes-Matopos National Park' Unpublished research paper, University of Manchester.

Ranger, T.O., 1989, 'Whose Heritage? The Case of the Matobo National Park' *Journal of Southern African Studies* 15: 217–50.

Raymond-Taylor, B. (ed.) 1995, *Ecological Resistance Movements: The Global Emergence of Radical and Popular Environmentalism* (New York: State University of New York).

Redclift, M., 1987, *Sustainable Development: Exploring the Contradictions* (London: Methuen).

Redclift, M., 1992, 'Sustainable Development and Global Environmental Change' *Global Environmental Change* 2: 32–42.

Reeve, R. and S. Ellis, 1995, 'An Insider's Account of the South African Security Forces Role in the Ivory Trade' *Journal of Contemporary African Studies* 13: 222–43.

Reno, W., 1995, *Corruption and State Politics in Sierra Leone* (Cambridge: Cambridge University Press).

Roe, E., 1992, *Report on the Amalgamation of District Councils and Rural Councils* CASS Occasional Papers Series NRM 7/1992 (Harare: CASS).

Rotberg, R., 1988, *The Founder: Cecil Rhodes and the Pursuit of Power* (New York: Oxford University Press).

Rothchild, D. and N. Chazan (eds) 1988, *The Precarious Balance: State and Society in Africa* (Boulder: Westview Press).

Rowell, A., 1997, 'Trouble in the Camp' *BBC Wildlife* 15: 34–6.

SACIM (Southern African Centre for Ivory Marketing), 1994, 'The SACIM Treaty: Its Aims and Objectives'. A Position Paper Presented by Botswana, Malawi, Namibia and Zimbabwe. Paper presented at the African Elephant in the Context of CITES Conference held in Kasane, Botswana, 19–23 September 1994 Department of Environment (DOE), Part B.

SADCC, 1988, *Natural Resources and the Environment: Policies and Development Strategy* (Maseru: SADCC).

Sagoff, M., 1994, 'Four Dogmas of Environmental Economics' *Environmental Values* 3: 285–310.

Sandbach, F., 1978, 'Ecology and the Limits to Growth Debate' *Antipode* 10: 22–32.

Sarre, P., 1995, 'Towards Global Environmental Values: Lessons from Western and Eastern Experience' *Global Environmental Values* 4: 115–27.

Save Valley Conservancy Trust, [1996?] *Proposal to Create an Interdependence Between the Surrounding Communities and the Save Valley Conservancy in the South-East Lowveld of Zimbabwe* (Chiredzi: SVC Trust).

Schoffeleers, J.M. (ed.), 1978, *Guardians of the Land: Essays on Central African Territorial Cults* (Gwelo, Zimbabwe: Mambo Press).

Schumacher, F., 1973, *Small is Beautiful* (London: Blond and Briggs).

Scoones, I., 1997, 'Landscapes, Fields and Soils: Understanding the History of Soil Fertility Management in Southern Zimbabwe' *Journal of Southern African Studies* 23: 615–34.

Scott, J.C., 1976, *The Moral Economy of the Peasant: Rebellion and Subsistence in South East Asia* (New Haven: Yale University Press).

Seligson, M.A. and J.T. Passe-Smith (eds), 1993, *Development and Underdevelopment: The Political Economy of Inequality* (London: Lynne Rienner).

Selous, F.C., 1893, *Travel and Adventure in South East Africa* (London: Rowland Ward).

Selous, F.C., 1896, *Sunshine and Storm in Rhodesia* (reprinted 1969: New York: Negro Universities Press).

Selwyn, T. (ed.), 1996, *The Tourist Image: Myths and Myth Making in Tourism* (Chichester: John Wiley and Sons).

Sen, G., 1995, 'Creating Common Ground Between Environmentalists and Women: Thinking Locally, Acting Globally?' *Ambio* 24: 64–5.

Sguazzin, A, 1995, 'SADC: Cross Border Parks Aimed to Aid Locals' *Africa Information Afrique, Southern Africa Chronicle* 8: 1.

Sheridan, E., 1994, 'Tales from the Bush: Getting Away from Cull' *BBC Wildlife* 12: 82.

Showers, K., 1989, 'Soil Erosion in the Kingdom of Lesotho: Origins and Colonial Response 1830s–1850s' *Journal of Southern African Studies* 15: 250–63.

Shrader-Frechette, K.S. and E.D. McCoy, 1994, 'How the Tail Wags the Dog: How Value Judgements Determine Ecological Science' *Environmental Values* 3: 107–20.

Simmons, I.G., 1993, *Interpreting Nature: Cultural Constructions of the Environment* (London: Routledge).

Sinclair, M.T., P. Alizadeh and E.A.A. Onunga, 1992, 'The Structure of International Tourism and Tourism Development in Kenya' in Harrison (ed.) *Tourism and the Less Developed Countries* pp. 47–63.

Singer, P., 1986, 'Animal Liberation' in Van De Veer and Pierce (eds) *People, Penguins and Plastic Trees* pp. 24–32.

Smiles, P., 1961, *A Game Ranger's Notebook* (London and Glasgow: Blackie).

Smith, A., 1776, *The Wealth of Nations* (reprinted 1970: London: Pelican Books).

Smith, J., 1994, 'Tourism and the Environment: A Zimbabwe Sun Perspective' *Southern African Economist* October. 23–5.

Smith, R.J., 1996, 'Sustainability and the Rationalisation of the Environment' *Environmental Politics* 5: 25–47.

Sonmez, S.F., 1998, 'Tourism, Terrorism and Political Instability' *Annals of Tourism Research* 25: 416–56.

Sonmez, S.F. and A.R. Graefe, 1998, 'The Influence of Terrorism Risk on Foreign Tourism Decisions' *Annals of Tourism Research* 25: 112–44.

Spector, B.I. and A.R. Korula, 1993, 'Problems in Ratifying International Environmental Agreements: Overcoming Initial Obstacles in the Post-agreement Negotiation Process' *Global Environmental Change* 3: 369–81.

Steele, P., 1995, 'Ecotourism: An Economic Analysis' *Journal of Sustainable Tourism* 3: 29–44.

Sterba, J.P., 1994, 'Reconciling Anthropocentric and Non-anthropocentric Environmental Ethics' *Environmental Values* 3: 229–44.

Stern, P.C., T. Dietz and L. Kalof, 1993, 'Value Orientations, Gender and Environmental Concern' *Environment and Behaviour* 25: 322–48.

Stern, P.C., T. Dietz and G.A. Guagano, 1995, 'The New Ecological Paradigm in Social-Psychological Context' *Environment and Behaviour* 27: 322–48.

Stoneman, C. and L. Cliffe, 1989, *Zimbabwe: Politics, Economics and Society* (London: Pinter).

Tandon, Y., 1995, 'Grassroots Resistance to Dominant Land-use Patterns in Southern Africa' in Raymond-Taylor (ed.) *Ecological Resistance Movements* pp. 161–76.

Tatham, G., 1986, '*The Rhino Conservation Strategy in the Zambezi Valley, Code Named Operation Stronghold'*. Unpublished policy document (Harare: DNPWLM).

Taylor, R.D., 1990, *Socio-economic Aspects of Meat Production from Impala Harvested in a Zimbabwean Communal Land.* WWF MAPS Project Paper No. 8 (Harare: WWF).

Taylor R.D., 1993, *Elephant Management in Nyaminyami District, Zimbabwe: Turning a Liability into an Asset.* WWF MAPS Project Paper No. 33 (Harare: WWF).

Taylor, R.D. and D.H.M. Cumming, 1993, *Elephant Management in Southern Africa.* WWF MAPS Project Paper No 40 (Harare: WWF).

Taylor, R.D., D.H.M. Cumming and C. Mackie, 1991, *Aerial Census of Elephant and other Large Herbivores in the Sebungwe.* WWF MAPS Project Paper No. 29. WWF Project No. 3749 (Harare: WWF).

Tesh, S.N., 1993, 'Environmentalism, Pre-environmentalism and Public Policy' *Policy Sciences* 26: 1–21.

Thomas, C., 1992, *The Environment in International Relations* (London: Royal Institute of International Affairs).

Thomas, K., 1983, *Man and the Natural World: Changing Attitudes in England 1500–1800* (Middlesex: Penguin).

Thomas, S.J., 1991, *The Legacy of Dualism and Decision-making: The Prospects for Local Institutional Development in Campfire.* Joint Working Paper Series Paper 1/92 (Harare: CASS, University of Zimbabwe and DNPWLM).

Thompson, J.L., 1983, 'Preservation of Wilderness and the Good Life' in Elliot and Gare (eds) *Environmental Philosophy* pp. 85–105.

Thomson, R., 1992, *The Wildlife Game* (Westville, RSA: The Nyala Wildlife Publications Trust).

Thornton, A. and D. Currey, 1991, *To Save an Elephant* (London: Bantam Books).

Thornton, A. and R. Reeves, 1992, 'The Spoils of War' *BBC Wildlife* 10: 24–8.

Tolba, M.K., 1995, 'Towards a Sustainable Development' *Ambio* 24: 66–7.

Traffic, 1992, *Traffic: Recommendations on Proposals to Amend the CITES Appendices at the 8th Meeting of the Conference of the Parties to CITES, Kyoto* (Cambridge: Traffic International).

Turner, R.K. (ed.), 1988, *Sustainable Environmental Management* (London: Belhaven Press).

Turton, B.J. and C.C. Mutambirwa, 1996, 'Air Transport Services and the Expansion of International Tourism in Zimbabwe' *Tourism Management* 17: 453–62.

Urry, J., 1992, 'The Tourist Gaze and the Environment' *Theory, Culture and Society* 9: 3–24.

Vail, L., 1976, 'Ecology and History: the Example of Eastern Zambia' *Journal of Southern African Studies* 3: 129–55.

Van De Veer, D. and C. Pierce (eds), 1986, *People, Penguins and Plastic Trees* (Belmont, California: Wadsworth).

Van Onselen, C., 1972, 'Reactions to Rinderpest in Southern Africa' *Journal of African History* 13: 473–88.

Vivian, J., 1994, 'NGOS and Sustainable Development in Zimbabwe: No Magic Bullets' *Development and Change* 25: 167–93.

Waller, R.D., 1990, 'The Tsetse Fly in Western Narok, Kenya' *Journal of African History* 31: 81–101.

Warner, S., M. Feinstein, R. Coppinger and E. Clemence, 1996, 'Global Population Growth and the Demise of Nature' *Environmental Values* 5: 285–301.

Warren, M.A., 1983, 'Rights of the Non Human World' in Elliot and Gare (eds) *Environmental Philosophy* pp. 109–34.

Warwick, D.P., 1993, 'The Politics and Ethics of Field Research' in Bulmer and Warwick (eds) *Social Research in Developing Countries* pp. 315–30.

Weaver, D. and K. Elliot, 1996, 'Spatial Patterns and Problems in Contemporary Namibian Tourism' *Geographical Journal* 162: 205–17.

Weiler, B. and C.M. Hall (eds) 1992, *Special Interest Tourism* (London: Belhaven).

Wells, M. et al., 1992, *People and Parks: Linking Protected Area Management with Local Communities* (Washington: World Bank/WWF/USAID).

Westfall, R.S., 1977, *The Construction of Modern Science* (Cambridge: Cambridge University Press).

White, G., 1789, *The Natural History of Selbourne* (reprint 1994: Oxford: Oxford University Press).

White, L., Jr, 1965, 'The Historical Roots of Our Environmental Crisis' *Science* 155: 1203–7.

Wight, P., 1994, 'Environmentally Responsible Marketing of Tourism' in Cater and Lowman (eds) *Ecotourism* pp. 39–55.

Wijnstekers, W., 1994, 'Basic Requirements for Trade in Parts and Derivatives of CITES Listed Species'. Paper presented at the African Elephant in the Context of CITES Conference held in Kasane, Botswana, 19–23 September 1994 Department of Environment (DOE) Part B.

Wildlife Conservation International, 1992, *The Ivory Trade and Conserving the African Elephant*. WCI Policy Report No. 1 (New York: WCI).

Wildlife Society of Zimbabwe, 1993, *Zimbabwe's Great Elephant Debate*. A Report on Discussions held during the AGM of the Wildlife Society of Zimbabwe at Hwange National Park on 13 August, 1993 (Harare: Wildlife Society).

Wildlife Society of Zimbabwe [1995a?] *Information*. Internal policy document (Harare: Wildlife Society).

Wildlife Society of Zimbabwe [1995b?] *Kuburi Wilderness Area: Management Plan and Conservation Policy*. Internal policy document (Harare: Wildlife Society).

Williamson, B.R., 1975, 'Seasonal Distribution of Elephant in Wankie National Park' *Arnoldia* 7.

Wilson K.B., 1989, 'Trees in Fields in Southern Zimbabwe' *Journal of Southern African Studies* 15: 369–83.

Winpenny, J., 1991, *Values for the Environment* (London: HMSO).

Woodhouse, P., 1992, 'Environmental Degradation and Sustainability' in Allen and Thomas (eds) *Poverty and Development in the 1990s* pp. 97–116.

Woods, M., 1999, 'A Burning Passion' *Geographical Magazine* March: 31–3.

World Bank, 1994, *World Development Report* (Oxford: Oxford University Press).

World Bank, 1996, *Corruption: A Major Barrier to Sound and Equitable Development* World Bank Brief (Washington DC: World Bank Group).

World Bank, 1997a, '*Corruption and Good Governance*' World Bank Group Issue Brief (Washington DC: World Bank).

World Bank, 1997b, '*Corruption: A Major Barrier to Sound and Equitable Development.*' (Washington DC: World Bank).

World Bank, 1997c, '*Reducing Corruption*' World Bank Policy and Research Bulletin 8 (Washington DC: World Bank).

World Bank/GEF [1997a?] *Medium Sized Projects* (Washington DC/New York: World Bank/UNDP/UNEP).

World Bank/GEF [1997b?] *Operational Guidance for Preparation and Approval of Medium Sized Projects* (Washington DC/New York: World Bank/UNDP/UNEP).

World Commission on Environment and Development, 1987, *Our Common Future* (Oxford: Oxford University Press).

World Tourism Organisation (WTO), UNEP and IUCN, 1992, *Guidelines: Development of National Parks and Protected Areas for Tourism* Technical Report Series No. 13 (WTO/UNEP).

Worster, D., 1987, *A History of Ecological Ideas* (Cambridge: Cambridge University Press).

WPA and CFU, 1992, *Conservation with Utilisation* (Harare: WPA/CFU).

Wunch, J.S. and D. Olowu (eds), 1990, *The Failure of the Centralised State* (Boulder: Westview Press).

WWF, 1970, *Wildlife in Crisis* (Gland, Switzerland, WWF-International)

WWF, 1982, *Yearbook* (Gland, Switzerland: WWF-International).

WWF, 1992a, *WWF-UK Guidelines for Answering Questions on Trophy Hunting.* Internal document (Godalming, Surrey: WWF-UK).

WWF, 1992b, *Help WWF Stop the Rhino Horn Trade* (Gland, Switzerland: WWF-International).

WWF, 1994a, *CITES and the African Elephant* (Gland, Switzerland: WWF-International).

WWF, 1994b, *WWF Position Statement on CITES 1994* (Gland, Switzerland: WWF-International).

WWF, undated, *CITES: What is it?* (Godalming, Surrey: WWF-UK).

Wynne, B., 1992, 'Uncertainty in Environmental Learning: Reconceiving Science and Policy in the Preventive Paradigm' *Global Environmental Change* 2: 111–27.

Yap, N., 1990, 'NGOs and Sustainable Development' *International Journal* 45: 75–106.

Young, O.R., 1989, 'The Politics of International Regime Formation: Managing Natural Resources and the Environment' *International Organisation* 43: 349–75.

Zambezi Society, 1993, *Environmental Advisory Service* (Harare: Zambezi Society).

Zimbabwe Tourism Development Corporation (ZTDC), 1993, *Proposed Marketing Policy and Strategy* (Harare: ZTDC).

Zimbabwe Trust, 1990, *Tenth Anniversary Report* (Harare: Zimbabwe Trust).

Zimbabwe Trust, 1992, *Wildlife: Relic of the Past or Resource of the Future? The Realities of Zimbabwe's Wildlife Policymaking and Management* (Harare: Zimbabwe Trust).

Zimbabwe Trust, 1994, *Natural Resources Management Project: Fourth Cycle Report of the Process Oriented Monitoring System* (Harare: Zimbabwe Trust).

Zimbabwe Trust, undated a, *Elephants and People: Partners in Conservation and Rural Development* (Harare: Zimtrust).

Zimbabwe Trust, undated b, *Zimtrust Strategic Support to Campfire* (Harare: Zimbabwe Trust).

Zimbabwe Trust, DNPWLM and the Campfire Association, 1990, *People, Wildlife and Natural Resources: The Campfire Approach to Rural Development* (Harare: WWF).

Zimmerman, M.E., 1994, *Contesting the Earth's Future: Radical Ecology and Post Modernity* (Berkeley: University of California Press).

Zimrights, 1995a, *The Welfare Organisations Amendment Bill 1995* (Harare: Zimrights).

Zimrights, 1995b, *Big Brother is Watching: Drastic New Powers of the Minister to Interfere with Private Voluntary Organisations* (Harare: Zimrights).

adaptive management 15
Adopt-A-Scout 117
Africa Resources Trust (ART) 14, 54, 106
African Wildlife Foundation (AWF) 163
aid conditionality 4, 122
Amnesty International 50, 64
Angola 59, 60, 130
animal liberation 6
animal rights 4, 14, 16, 18, 50, 76, 104, 105, 112, 115, 120, 129, 133, 139, 157, 161, 171, 174
anti-poaching 46, 174, 176
    funding 51, 52, 58, 123
    incentive scheme 51
    indemnity 50
    military support (Special Support Units) 54
    NGO involvement 52, 54
    patrols 50, 51
    shoot-to-kill 49, 50, 65
    war on poachers 48
appropriate authority 16, 75, 90, 107, 121

Beit Trust 83
Black Rhino Conservation Strategy 47, 55, 80, 82
blue-greens 5, 86, 88
Bophutatswana 133
Botswana 16, 47, 72, 130, 156, 157, 164, 166, 167, 170
Bruntland Commission 4
Bubiana Conservancy 78, 88, 132

Cabinet Development Committee 28, 32
Campfire Association 92, 100, 106, 111, 170
Campfire Collaborative Group 106, 107
Campfire Coordination Unit 109
Care for the Wild 36, 124,125, 133, 139
cattle barons 95, 96
Centre for Applied Social Science (CASS) 106
Central Intelligence Organisation 40, 64
Chimutengwende, Chen 36, 37, 85, 168
Chinese medicine (traditional) 149
Colonialism 10, 11
commercial farmers 69, 83, 86, 98, 175, 176
    beef production 75, 175
Commercial Farmers Union (CFU) 77
common property regimes 91, 93
Communal Areas Management Plan for Indigenous Resources (Campfire) 2, 3, 15, 45, 74, 89, 116, 174
    Central government 110, 111
    communal property 92
    District Councils 93, 94, 98, 99, 107, 110, 112
    donor involvement 135
    empowerment 96
    gender 95
    hunters 100, 102, 103
    ivory trade 101, 104, 105, 112, 155, 170

revenue distribution 98–101, 109
squatters 94
subsistence poaching 97, 101
sustainable development 91
tourism 91
traditional beliefs 92
wildlife versus people 97, 98
conservancies 33, 78, 88, 175
anti-poaching 80
black rhino conservation 81, 83
community interests 80
financial viability 79
poaching 83
Price Waterhouse Study 79
conservation 1, 13
consumptive use 17, 77, 150, 151, 177
Convention on the International Trade in Endangered Species (CITES) 3, 21, 31, 38, 45, 106, 120, 131, 173
Appendix listings 143, 144, 157
enforcement 146, 147
prohibition regimes 141, 145, 152, 154, 171
regulatory regimes 141
reservations 165
split listings 143, 164, 165
Criminal Investigations Branch (police) 36
crooks corner 83, 138
cropping, wildlife 77, 112, 175
culling 18, 53, 130, 133, 139, 155, 156

David Shepherd Foundation 125
deep-greens, deep ecology 6, 12, 121, 129, 151, 161, 172, 176
dehorning (rhinos) 55, 81
Department for International Development (DFID) 134
Department of National Parks and Wildlife Management (DNPWLM), (*see* Parks Department)
Deutsche Gesellschaft für Technische Zusammerarbeit (GTZ) 36, 133
Development Trust of Zimbabwe (DTZ) 39
District Councils 16, 90, 93, 94, 98, 99, 107, 110, 121
decentralisation 108
recentralisation 108, 109
donors 133, 136

ecofeminists 6
economics of extinction 153
Economic Structural Adjustment Plan (ESAP) 25, 26, 109
ecotourism 71
elephant counting 131
elephant communication 132
elephant export scandal 34, 38
Elephant Rose Foundation 115, 117
Endangered Species Act (USA) 41, 146
environmental economics 16, 68, 72, 73, 150
environmental imperialism 115, 122, 139, 155, 176
Environmental Investigation Agency (EIA) 59, 124, 126, 129, 139, 147, 162, 163, 176

Farm Management Africa 99
fencing (electric) 98
Frente de Libertação de Moçambique (Frelimo) 62
Friends of Animals 163
Friends of the Earth 123

Gaia Hypothesis 6
Global Environmental Facility (GEF) 136, 170
globalisation 74, 122, 155
Gonarezhou Drought Crisis Committee 35
Gonarezhou National Park 24, 34, 38, 60, 63, 64, 72, 82, 131, 132
Greenpeace 113, 123, 126, 167
group access tenure 92
Guruve District 90, 101, 104, 111

Hong Kong 148, 157, 158, 166
Humane Society of America 129, 130, 176
hunters 10
hunting, sport 18, 72, 75, 77, 100, 103, 104, 112, 124, 127, 130, 150
hunting, subsistence, 11, 24
Hwange National Park 25, 26, 47, 55, 131, 133, 134

Independence 12
indigenisation 12, 31, 33, 86, 107
International Black Rhino Foundation 52

International Fund for Animal Welfare (IFAW) 133, 167
International Monetary Fund 70
Intensive Protection Zones (IPZs) 55, 56
International Union for the Conservation of Nature (IUCN) 57, 127, 145
International Criminal Police Organisation (Interpol) 143
Investigations Branch (Parks Department) 36, 40
IUCN Elephant Specialist Group 57, 169
Ivory Export Quota System 159
ivory trade 38, 104, 105, 141,152
   Appendix I listing 157, 168
   Appendix II listing 157, 167, 170
   anti-ban states 154, 161, 165, 168–169
   carving industry 158, 160
   consuming states 152, 158
   illegal trade 58, 59, 158
   legal trade 157, 159
   post-ban 166, 171
   pro-ban states 153, 154, 161, 165, 168, 169
   seizures 159, 169
   stockpiles 105, 154, 159, 169
   trade ban 3, 57, 142, 153, 157, 160, 166
Ivory Trade Review Group 165

Japan 148, 154, 157, 158, 167, 171
Japan Ivory Importers Association 158

KAS Enterprises 60, 61
Kenya 48, 67, 72, 148, 153, 155, 156, 158, 160, 166, 169
Kenya Wildlife Service 153, 158
Kruger National Park 82, 137
Kumleben Commission 47, 59, 61, 65

Lacey Act (USA) 146
Lancaster House Agreement 11
Land Acquisition Act, 1992 78, 84, 85
land question 10, 13, 84, 85, 88, 91, 145, 176, 177
Leakey, Richard 153, 158
liberation war 11
Limits to Growth 3

Mafia 148, 160
Makombe, Willis 28, 31, 33, 37
Mana Pools National Park 117, 131
Martin, Rowan 15, 28, 29, 31, 33, 35, 37, 61, 138, 145, 162
Midlands Conservancy 78, 88, 124
Military Intelligence Department (South Africa) 59
Ministry of Environment and Tourism 13, 22, 25, 28, 34, 35, 40, 53, 84, 88, 118, 135, 175
Ministry of Finance 28
Ministry of Local Government, Rural and Urban Development (MLGRUD) 108, 110, 111, 112
Mozambique 16, 54, 58, 59–61, 62, 64, 82, 130, 137
Mozambique National Resistance (*see* Renamo)
Multi Species Animal Production Systems (MAPS) 106, 127
Mundangepfupfu, Tichafa 28, 31, 40
Murerwa, Herbert 31, 138

Namibia 47, 60, 75, 156, 166, 167, 169
National Conservation Strategy 12
National Parks 10, 22
   frozen zones 62, 63
   principle of exclusion 23, 89, 174
Nduku, Willie 28, 29, 31, 33, 35, 37, 138
Nkomo, Joshua 28, 31, 38, 39, 110, 112, 175
Nleya, Captain 64
non-consumptive use 17, 77, 102, 130, 151, 177
Non Governmental Organisations (NGOs)
   campaigning 126, 130
   funds 122, 123
   international NGOs 120, 126, 161, 165
   lobbying 117
   professionalisation 113
   Zimbabwean NGOs 115, 119
Nyaminyami 90, 98
Nyaminyami Wildlife Management Trust 97

open access tenure 92
Operation Lock 60, 61

Operation Oasis 133
Operation Safeguard Our Heritage
(Operation Save Our Heritage) 54,
55, 65
Operation Stronghold 48, 51, 53, 65
Overseas Development Administration
(ODA) 134, 135
Oxfam 115

Pangeti, George 29, 37, 138
Parks Department 9, 13, 21, 49
brain drain 26, 32
Campfire, 105, 106
commercialisation 27, 29, 138, 175
conservationist alliance 31, 69, 175
financial crisis 25, 27, 69, 87, 122,
134, 168
institutional capacity 22
poaching faction 38, 63
patronage networks 28, 34
patronage alliance 31
power struggle 29, 38, 69, 175, 177
privatisation 27
Parks and Wildlife Act 1975 16, 23, 68,
75, 78, 85, 90, 108, 121
Parks and Wildlife Estate 22, 27, 29,
39, 55, 67, 138, 175
patron-client networks 28, 34
poachers 44, 45
poaching 38, 139, 166, 177
commercial poaching 44, 45, 56, 176
definition of 43
diplomatic community 47, 160
human rights violations 48
ivory poaching 56, 66
military involvement 54, 56, 60, 61
official involvement 58, 61, 63
rhino poaching 43, 56, 124
Rhodesian Army 62
subsistence poaching 44, 45, 97, 174
trucking companies 47, 63
policy slippage 30
'politics of the belly' 30
precautionary principle 14, 152, 153
preservation 1, 2, 14, 23, 48, 55, 112,
115, 120, 121, 161
Private Voluntary Organisations
(PVOs) 118
privatisation (of international
relations) 52, 122, 176

privatisation (of wildlife) 67
Problem Animal Control (PAC) 98,
103, 155
Protection of Wildlife (Indemnity) Act
50
Putterill, Gordon 34, 38

red-greens 5, 12
Renamo 49, 59, 60, 62, 65
rhino horn trade 38, 57, 124, 141
aphrodisiac myth 149
consuming states 149
illegal trade 147, 150
sport hunting 150, 151
stockpiles 150
Taiwanese traders 148
trade ban 3, 45, 150, 151, 171
Rhino Rescue 125, 139
rhino wars/poaching 44, 47, 52
Roos Commission 59
Rukuni Commission, 1994 96, 108
rumour 30, 31
Rupert, Anton 128

SAVE 117
Save Our Wildlife Heritage 133
Save Valley Conservancy 78, 88, 132,
137
Save Valley Wildlife Services Ltd. 80
science, ideology of 2
Sebakwe Black Rhino Trust 123, 124,
125, 139
settler agriculture 10, 12
Sithole, Reverend 33
smuggling 58, 61, 63, 64, 65, 159, 169
Social Welfare Organisations
Amendment Act, 1995 118
South Africa 54, 58, 72, 75, 78, 157
destabilisation policy 59, 61, 63, 65
South African Defence Force (SADF)
47, 59, 61, 62
Southern African Centre for Ivory
Marketing (SACIM) 156
Southern African Convention for
Wildlife Management (SACWM)
156, 157
Southern African Development
Community (SADC) 169, 171
Southern African Group for the
Environment (SAGE) 115, 117, 151

Southern Rhodesia 10
Species Survival Network 167
squatters 94
Superpark 82, 136, 139
sustainability, definition of 4, 5
sustainable development 4, 5
sustainable utilisation 9, 15, 18, 116,
    120, 140, 174, 177
Swaziland 82

Taiwan 150
Tanzania 67, 130, 148, 153, 160, 164,
    166
torture 37, 64
tourism industry 11, 23, 25, 68, 70, 74,
    87, 102, 103, 131, 136
  ecotourism 71, 102
  phototourism 77, 78, 102
tourists 76
Trade Records Analysis of Flora and
    Fauna in International Commerce
    (Traffic) 143, 169
traditional leaders 94
tragedy of the commons 77, 92
Transfrontier Conservation Areas
    (TFCAs) (*see* Superparks)
translocation
  elephants 34, 38, 132, 133
  rhinos 48
Tusk 36, 123, 125, 133

União Nacional para a Independência
    Total de Angola (UNITA) 59, 65
United Nations Conference on
    Environment and Development
    (UNCED) 3, 136
United Nations Environment
    Programme (UNEP) 136
USAID 124, 135
US Fish and Wildlife Service 36, 146

Village Development Committees
    (VIDCOs) 108
VIP hunting scandal 38, 41

Welfare Advisory Board 119
white farmers 38, 116, 118
wildlife auctions 76, 77
Wildlife Management Services 35, 36,
    133

Wildlife Producers Association (WPA)
    35, 77
wildlife ranching 74, 78, 86, 87
Wildlife Society 86, 117, 122
wildlife utilisation 13, 14
World Bank 4, 25–27, 70, 115, 133,
    135, 162
World Wide Fund for Nature (*see*
    World Wildlife Fund)
World Wildlife Fund (WWF)
    International 53, 60, 113, 126,145,
    147
  1001 Club 129, 130, 164
  anti-poaching/shoot-to-kill 53
  conservancies 83
  funders 129
  ivory ban 162, 165
  splits 126, 128
World Wildlife Fund – South Africa 129
World Wildlife Fund – UK 124, 127,
    128, 164
World Wildlife Fund – US 127, 164,
World Wildlife Fund-Zimbabwe 75,
    106, 118

Yakuza 148
Year of the Elephant 1988 163

Zambezi Society 56, 58, 117, 118
Zambezi valley 39, 46, 48, 51, 55, 56,
    57, 81, 82
Zambia 16, 46, 47, 49, 50, 57, 58, 130,
    164
Zimbabwe African National Union
    (ZANU) 11
Zimbabwe African National Union-
    Ndonga (ZANU-Ndonga) 33
Zimbabwe African National Union-
    Patriotic Front (ZANU-PF) 28, 32,
    50, 111, 119, 137, 17
Zimbabwe Association of Tour and
    Safari Operators (ZATSO) 103
Zimbabwe Council for Tourism 72
Zimbabwe Hunters Association 40, 76
Zimbabwe National Army (ZNA) 61,
    63, 65, 176
Zimbabwe Sun Group 72
Zimbabwe Tourism Development
    Corporation (ZTDC) 71
Zimbabwe Trust 36, 106